AF767218

MATHEMATICAL LOGIC, COMPUTABILITY, COMPLEXITY, AND RANDOMNESS

MATHEMATICAL LOGIC, COMPUTABILITY, COMPLEXITY, AND RANDOMNESS

16th Asian Logic Conference and 14th International Conference on Computability, Complexity, and Randomness

Nazarbayev University, Astana, Kazakhstan, 17 – 21 June 2019 and 23 – 25 June 2019

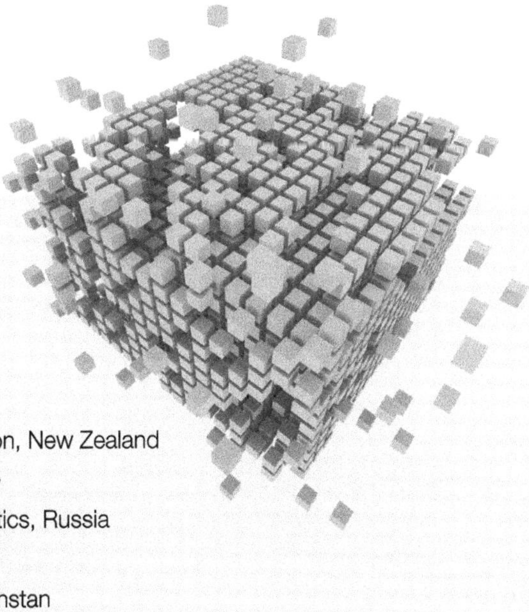

Editors

Joerg Brendle
Kobe University, Japan

Rod Downey
Victoria University of Wellington, New Zealand

Sergey S. Goncharov
Sobolev Institute of Mathematics, Russia

Manat Mustafa
Nazarbayev University, Kazakhstan

Guohua Wu
Nanyang Technological University, Singapore

Yue Yang
National University of Singapore, Singapore

World Scientific

NEW JERSEY • LONDON • SINGAPORE • BEIJING • SHANGHAI • HONG KONG • TAIPEI • CHENNAI • TOKYO

Published by

World Scientific Publishing Co. Pte. Ltd.
5 Toh Tuck Link, Singapore 596224
USA office: 27 Warren Street, Suite 401-402, Hackensack, NJ 07601
UK office: 57 Shelton Street, Covent Garden, London WC2H 9HE

British Library Cataloguing-in-Publication Data
A catalogue record for this book is available from the British Library.

MATHEMATICAL LOGIC, COMPUTABILITY, COMPLEXITY, AND RANDOMNESS
Proceedings of the 16th Asian Logic Conference and 14th International Conference on Computability, Complexity, and Randomness

Copyright © 2026 by World Scientific Publishing Co. Pte. Ltd.

All rights reserved. This book, or parts thereof, may not be reproduced in any form or by any means, electronic or mechanical, including photocopying, recording or any information storage and retrieval system now known or to be invented, without written permission from the publisher.

For photocopying of material in this volume, please pay a copying fee through the Copyright Clearance Center, Inc., 222 Rosewood Drive, Danvers, MA 01923, USA. In this case permission to photocopy is not required from the publisher.

ISBN 978-981-98-1593-7 (hardcover)
ISBN 978-981-98-1594-4 (ebook for institutions)
ISBN 978-981-98-1595-1 (ebook for individuals)

For any available supplementary material, please visit
https://www.worldscientific.com/worldscibooks/10.1142/14386#t=suppl

Preface

The *Sixteenth Asian Logic Conference (ALC 2019)* was held at Nazarbayev University, Astana, Kazakhstan, during June 17-21, 2019.

ALC 2019 was the second Asian Logic Conference since its status changed from an ASL (the Association for Symbolic Logic) sponsored meeting to an official ASL meeting by ASL Council action in May 2016.

The Asian Logic Conference (ALC) is a major international event for mathematical logic. It features the latest scientific developments in the fields of mathematical logic and its applications, logic in computer science, and philosophical logic. The ALC series also aims to promote mathematical logic in the Asia-Pacific region and to bring logicians together, from within Asia as well as from the rest of the world to exchange information and ideas. From 1981 to 2008, the Asian Logic Conference has been held triennially and by rotation of countries in the Asia-Pacific region. The previous meetings took place in Singapore (1981), Bangkok (1984), Beijing (1987), Tokyo (1990), Singapore (1993), Beijing (1996), Hsi-Tou (1999), Chongqing (2002), Novosibirsk (2005), and Kobe (2008). In 2008, the East Asian and Australasian Committees of the Association for Symbolic Logic decided to shorten the three-year cycle to two. The new two-year cycle began with the meeting in Singapore (2009), and subsequent meetings have been held in Wellington (2011), Guangzhou (2013), Mumbai (2015), Daejeon (2017).

ALC 2019 was co-located with CCR 2019, the 14th International Conference on Computability and Randomness. Plenary keynote lectures were given by Serikzhan Badaev (Al-Farabi Kazakh National University, Kazakhstan), Nikolay Bazhenov (Sobolev Institute of Mathematics, Russia), Qi Feng (Chinese Academy of Sciences, China), Sakae Fuchino (Kobe University, Japan), Takayuki Kihara (Nagoya University, Japan), Byunghan Kim (Yonsei University, Korea), Beibut Kulpeshov (IIT University, Kazakhstan), Dietrich Kuske (Technische Universität Ilmenau, Germany), Dugald Macpherson (University of Leeds, UK), Ahti-Veikko Pietarinen (Nazarbayev University, Kazakhstan), Zhiwei Sun (Nanjing University, China), and Toshimichi Usuba (Waseda University, Japan). Fifteen invited lecturers gave special sessions on Computability theory, Model theory, Set theory, Logic in Computer Science and Philosophical Logic.

We have kept our tradition of choosing World Scientific as the publisher of ALC proceedings, and in this volume, 8 refereed papers of speakers are presented. We express our thanks to all who contributed to this volume, especially to authors and reviewers.

Editors

Joerg Brendle, Rod Downey, Sergey S. Goncharov,
Manat Mustafa, Guohua Wu, Yue Yang

Organizing committees

Programme Committee of ALC 2019:

Sergey S. Goncharov (Chair)	– Sobolev Institute of Mathematics, Russia
Bektur Baizhanov	– Institute of Mathematics, Kazakhstan
Jörg Brendle	– Kobe University, Japan
Koichiro Ikeda	– Hosei University, Japan
Bakhadyr Khoussainov	– University of Auckland, New Zealand
Andrey Morozov	– Sobolev Institute of Mathematics, Russia
Hiroakira Ono	– Advanced Institute of Science and Technology, Japan
Wei Wang	– Sun Yat-sen University, China
Guohua Wu	– Nanyang Technological University, Singapore
Yue Yang	– National University of Singapore, Singapore
Liang Yu	– Nanjing University, China

Programme Committee of CCR 2019:

Rod Downey (Chair)	– Victoria University of Wellington, New Zealand
Yue Yang (Co-Chair)	– National University of Singapore, Singapore
Bruno Bauwens	– National Research University Higher School of Economics, Russia
Daniel Turetsky	– University of Notre Dame, USA
Serikzhan Badaev	– Al-Farabi Kazakh National University, Kazakhstan
Keng Meng (Selwyn) Ng	– Nanyang Technological University, Singapore
George Barmpalias	– Chinese Academy of Sciences, China
Johanna N. Y. Franklin	– Hofstra University, USA
Barbara F. Csima	– University of Waterloo, Canada
Denis Hirschfeldt	– University of Chicago, USA

Local Committee:

Manat Mustafa (Chair)	– Nazarbayev University, Kazakhstan
Vassilios D. Tourassis (Co-Chair)	– Nazarbayev University, Kazakhstan
H. Atakan Varol (Co-Chair)	– Nazarbayev University, Kazakhstan
R. Zhumanova	– Nazarbayev University, Kazakhstan
Zh. Kadyrsizova	– Nazarbayev University, Kazakhstan
F. Sica	– Nazarbayev University, Kazakhstan

Contents

Comparing the isomorphism types of equivalence structures and preorders

Nikolay Bazhenov

Sobolev Institute of Mathematics, pr. Akad. Koptyuga 4, Novosibirsk, 630090, Russia
E-mail: bazhenov@math.nsc.ru

Luca San Mauro

Institute of Discrete Mathematics and Geometry, Vienna University of Technology, Vienna, Austria
E-mail: luca.san.mauro@tuwien.ac.at

A general theme of computable structure theory is to investigate when structures have copies of a given complexity Γ. We discuss such a problem for the case of equivalence structures and preorders. We show that there is a Π^0_1 equivalence structure with no Σ^0_1 copy, and in fact that the isomorphism types realized by the Π^0_1 equivalence structures coincide with those realized by the Δ^0_2 equivalence structures. We also construct a Σ^0_1 preorder with no Π^0_1 copy.

Keywords: Equivalence structures, preorders, Ershov hierarchy, arithmetical hierarchy, analytical hierarchy, isomorphism types.

1. Introduction

A primary question of computable structure theory is to ask, given a familiar class of countable structures $\mathfrak{K}$, which of the isomorphism types of $\mathfrak{K}$ can be realized by computable structures. Sometimes the answer is trivial: for instance, *all* countable vector spaces over $\mathbb{Q}$ have a computable copy[1]; on the other hand, Peano Arithmetic has only *one* computable model[2]. In other cases, such as abelian p-groups, one (partially) answers in terms of a class of invariants[3]. But many natural classes (e.g., linear orders, Boolean algebras, or the class of all graphs) are simply too rich to hope for a nice characterization of which members of the class have a computable copy.

An obvious way of generalizing the above problem is to investigate when structures have copies of some complexity Γ — where Γ is some level of the arithmetical, hyperarithmetical, analytical, or Ershov hierarchy — and then studying the relation between the isomorphism types realized by structures of different complexity. Interestingly, it can be the case that all the isomorphism types realized by structures of a certain complexity are already realized by structures of considerably less complexity: e.g., Spector famously proved that every hyperarithmetic ordinal has a computable copy[4]. In this paper, we contribute to this research thread by focusing on two case-studies: equivalence structures and preorders.

Equivalence structures are among the simplest structures which are nontrivial. For this reason, they form a class which is often reasonably tame but still worth examining. In fact, despite of their structural simplicity, equivalence structures have also remarkably deep effective properties: Calvert, Cenzer, Harizanov, and Morozov[5] proved that the index set of the computable equivalence structures isomorphic to a given one may be Π^0_4-complete; Downey, Melnikov, and Ng showed[6] that there is an effective list of all computable equivalence structures, thus classifying them in a sense proposed by Goncharov and Knight[7]. Cenzer, Harizanov, and Remmel[8] started the study of the isomorphism types of Σ^0_1 and Π^0_1 equivalence structures; our work here extends their study.

Another source of motivation for focusing on equivalence structures comes from the fact that measuring the complexity of equivalence relations with domain ω has been a longstanding endeavour in the literature. In particular, for decades, researchers investigated the effective properties of computably enumerable equivalence relations (called *ceers*, after Gao and Gerdes[9]) and they compared them via a natural effectivization of Borel reducibility called *computable reducibility*: see, e.g.,[10–14]. In comparison with ceers, the study of equivalence relations of higher complexity has been less extensive. Yet, computable reducibility has been used to analyze equivalence relations of various complexity, including some that are not even hyperarithmetical (such as the isomorphism relations for familiar classes of computable structures[15]). In recent times, following the work of Ng and Yu[16], we initiated a systematic study of Δ^0_2 equivalence relations: we proved that theory of ceers, co-ceers, and Δ^0_2 equivalence relations behave quite differently[17]. This motivates the following question (from which the present paper originated):

Are ceers and co-ceers distinguishable in terms of the isomorphism types that they realize?

Theorem 2.1 below positively answers to such question. In fact, we give a full description of the comparison between the isomorphism types realized at different levels of the arithmetical hierarchy: in particular, we show that for a non-zero n, the isomorphism types of Δ^0_{n+1} equivalence structures are precisely the same as those of Π^0_n equivalence structures (Theorems 2.2 and 2.3). Furthermore, we separate different levels of the analytical hierarchy via the corresponding realizable types of equivalence structures (Proposition 2.1).

Our second case-study is also inspired by the research on computable reducibility. Preorders occur naturally in logic (e.g., think of the structure induced by $\vdash_{PA}$ on the formulas of arithmetic) and, in fact, their complexity has been studied since the 1980s[18] (see[19] for more recent results). By Theorem 3.1 below we show that the case of preorders differs from that of the equivalence structures.

1.1. *Preliminaries on equivalence structures and preorders*

Equivalence structures (resp. preorders) are structures of the form $\mathcal{A} = \langle \omega, R \rangle$ where R is an equivalence relation (a preorder, i.e., a reflexive and transitive relation); $\mathcal{A} \in \Sigma^i_\alpha$, for $i \in \{-1, 0, 1\}$, if $R \in \Sigma^i_\alpha$. Note that, for our interests, assuming that all structures have domain ω is not a limitation: every isomorphism type of an equivalence structure with a countably infinite domain is realized by an equivalence structure with domain ω.

We denote by $\mathbb{EQ}$ and by $\mathbb{PO}$ the class of all countable equivalence structures and all countable preorders respectively. Moreover, for any class of structures $\mathbb{K}$, we denote by $\mathrm{IsoType}(\mathbb{K}, \Gamma)$ the collection of all isomorphism types realized by the structures from $\mathbb{K}$ of complexity Γ.

We denote by $\{R_e\}_{e \in \omega}$ some fixed uniform enumeration of all ceers. We say that an equivalence class $[x]_R$ is *older* than an equivalence class $[y]_R$ if $\min[x]_R < \min[y]_R$. This relation obviously induces a linear ordering on the R-classes.

1.2. *Warning about our terminology*

There is a well-established way of thinking of a Σ^0_1 (resp. Π^0_1) *presentation* of structure $\mathcal{A}$ as the combination of a pre-structure $\mathcal{A}^*$ and a ceer (co-ceer) R such that the quotient structure $\mathcal{A}^*/R$ is isomorphic to $\mathcal{A}$. This approach generalizes the way in which many algebraic structures are presented (e.g., the homomorphism theorem says that every countable group is a quotient of the free group on countably many generators). The study of quotient presentations dates back to the very beginning of computable model theory (see, e.g., the work of Metakides and Nerode on c.e. vector spaces[20]) and has given rise to a rich research program; the interested reader is referred to[21–23]. However, it should be clear that in this paper we do *not* deal with quotients, but rather with the (pre-)structure themselves. We emphasize this point with one example. Let R be a preorder on ω. One can regard at the algebraic properties of R in two ways:

(a) First, considering the quotient structure $\mathcal{A} := \langle R/supp(R), \leq_R \rangle$, where

- $supp(R) := \{(x, y) : (x \; R \; y) \& (y \; R \; x)\}$;
- the domain of $\mathcal{A}$ consists of the equivalence classes $[x]_{supp(R)}$, $x \in \omega$;
- and $[x]_{supp(R)} \leq_R [y]_{supp(R)}$ if and only if $(x \; R \; y)$.

(b) Secondly, considering directly the structure $\langle \omega, R \rangle$.

Feiner's celebrated result[24] that there is Σ^0_1-*presented* linear ordering which is not isomorphic to any computable one shall be understood within the first setting. On the contrary, in the present paper we will always work according to the second setting. For example, if we say that R is a Σ^0_1 *linear order*, then this means the following:

- $R \in \Sigma^0_1$, and
- the ordering R is linear (in particular, R is an antisymmetric relation).

Clearly, these conventions imply that in our setting, *every* Σ^0_1 linear order is computable.

1.3. *Preliminaries on limitwise monotonicity*

The system of invariants which characterize the isomorphism types of computable equivalence structures can be nicely represented in terms of limitwise monotonic functions.

Let $\mathbf{d}$ be a Turing degree. A function $F\colon \omega \to \omega \cup \{\infty\}$ is $\mathbf{d}$-*limitwise monotonic* if there is a total $\mathbf{d}$-computable function $f(x, s)$ such that

(a) $f(x, s) \leq f(x, s+1)$ for all x and s,
(b) $F(x) = \lim_s f(x, s)$ for all x.

When one is talking about $\mathbf{0}$-limitwise monotonic functions, the prefix "$\mathbf{0}$-" is typically omitted.

A set $A \subseteq \omega$ is *limitwise monotonic* if either $A = \emptyset$ or there is a limitwise monotonic function $F(x)$ with $range(F) = A$.

Theorem 1.1 (Calvert, Cenzer, Harizanov, and Morozov[5]). *Let $\mathcal{A}$ be a countable equivalence structure with infinitely many classes. Then the following conditions are equivalent:*

(a) *$\mathcal{A}$ has a computable copy.*
(b) *There is a limitwise monotonic function $F\colon \omega \to \omega \cup \{\infty\}$ such that for any $\kappa \in \omega \cup \{\infty\}$, the structure $\mathcal{A}$ has precisely $card(\{x \in \omega : F(x) = \kappa\})$ classes of size κ.*

We conclude this brief discussion about limitwise monotonicity by stating two results that we will use later. The reader is referred to, e.g.,[25] for further results on limitwise monotonic sets and functions.

Theorem 1.2 (Harris[26] and, independently, Kach[27]). *A function $F\colon \omega \to \omega \cup \{\infty\}$ is $\mathbf{0}'$-limitwise monotonic if and only if there exists a computable function $g\colon \omega \times \omega \to \omega$ such that*

$$F(x) = \liminf_s g(x, s) \text{ for all } x.$$

Theorem 1.3 (Khoussainov, Nies, and Shore[28]). *There is a d.c.e. set $A \subset \omega$, which is not limitwise monotonic.*

2. Isomorphism types of equivalence structures

Cenzer, Harizanov, and Remmel[8] proved that there is a Σ^0_1 equivalence structure with no computable copy. It is then natural to ask whether the Σ^0_1 and the Π^0_1 equivalence structures realize the same isomorphism types. The next theorem shows that this is not the case: Π^0_1 equivalence structures are, in a sense, more expressive that Σ^0_1 equivalence structures.

Theorem 2.1. *There is a Π^0_1 equivalence structure $\mathcal{A}$ which is not isomorphic to any Σ^0_1 equivalence structure. That is,* $\mathrm{IsoType}(\mathbb{EQ}, \Pi^0_1) \not\subseteq \mathrm{IsoType}(\mathbb{EQ}, \Sigma^0_1)$.

Proof. To prove the theorem, it is enough to build a co-ceer that is not isomorphic to any ceer. To do so, we construct in stages a co-ceer S which, for all e, satisfies the following requirements:

$$P_e : \text{There is an } S\text{-class of size } e+1 \Leftrightarrow \text{There are no } R_e\text{-classes of size } e+1.$$

This obviously guarantees that S cannot be isomorphic to any of the R_e's.

The construction

In the construction we will define the set Y_e of *e-witnesses*; at any stage, the number of e-witnesses will be either $e+1$ or $e+2$. The role of such e-witnesses will be that of realizing an equivalence class (i.e., $[\langle e,0\rangle]_S$) which will eventually ensure that $S \not\cong R_e$. During the construction we will often say that we *transform some z into a S-singleton*; by this we mean that we let $y \not{S} z$ and $z \not{S} y$, for all $z \neq y$. Finally, we say that P_e *turns* ***on*** at some stage s if the oldest $R_e[s]$-class of size $e+1$ differs, for all $t < s$, from the oldest $R_e[t]$-class of the same size.

Stage 0

For all e, let $Y_e[0] := \{x : 0 < x \leq e+1\}$ and let all the requirements P_e be **off**. Moreover, let $S[0]$ be the following computable partition of ω in infinitely many classes:

$$\langle a,b\rangle S[0]\langle c,d\rangle \Leftrightarrow a = c.$$

Stage $s+1 = \langle e,n\rangle$

We focus on P_e. First, let

$$u_e := \min\{x : x \in Y_e[s] \text{ and } x > e+1\}$$

(if such set is not empty), and let

$$v_e := \min\{x : x > \max Y_e[s] \text{ and } \langle e,0\rangle S[s]\langle e,x\rangle\}.$$

Now, we distinguish four cases:

(1) $card(Y_e[s]) = e+1$, $R_e[s+1]$ has some class of size $e+1$, and P_e is **off**: If so, we want to declare a new e-witness. We do so by setting $Y_e[s+1] := Y_e[s] \cup \{v_e\}$. We also transform $\langle e, v_e+1\rangle$ into a S-singleton;

(2) $card(Y_e[s]) = e+2$, $R_e[t]$ has no classes of size $e+1$, and P_e is **off**: If so, we want to decrease the number of e-witnesses by 1. We do so by setting $Y_e[s+1] := Y_e[s] \smallsetminus \{u_e\}$ and transforming $\langle e,u_e\rangle$ into an S-singleton;

(3) If P_e is **on**: If so, we want to change the current set of e-witnesses. We do so by setting $Y_e[s+1] := (Y_e[s] \smallsetminus \{u_e\}) \cup \{v_e\}$ and transforming $\langle e,u_e\rangle$ into an S-singleton. We also turn P_e **off**;

(4) Any other case: We transform $\langle e,v_e\rangle$ into an S-singleton.

Finally, let $Y_e := \lim_{s\to\infty} Y_e[s]$ and let $S := \bigcap_{s\in\omega} S[s]$.

The verification

By construction, S is obviously a co-ceer. Moreover, by induction it is not difficult to see that, for all e, $Y_e[0] \subseteq Y_e$ and the following identity holds:

$$card([\langle e,0\rangle]_S) = card(Y_e). \qquad (\star)$$

The rest of the verification is based on the following lemma.

Lemma 2.1. *For all e, $S \not\geq R_e$.*

Proof. Suppose first that R_e has an equivalence class of size $e+1$. So, there exists a stage s_0 such that any $R_e[s]$, with $s \geq s_0$, realizes the oldest equivalence class of size $e+1$ of R_e. Let t be the least number such that $\langle e,t\rangle \geq s_0$. It is not difficult to see that, since s_0 is chosen minimal, at stage $\langle e,t\rangle$ we enter in case (1) of the construction, by which we set the number of e-witnesses to $e+2$. Next, observe that in all further stages in which we focus on P_e we always enter in case (4) (case (1) is excluded because $card(Y_e) \neq e+1$; case (2) is excluded because R_e has some class of size $e+1$; case (3) is excluded because $R[s_0]$ already realized the oldest R-class of size $e+1$ and therefore P_e stays **off**). This means that we will not modify the set of e-witnesses. Therefore, we have that $Y_e = Y_e[\langle e,t\rangle]$ and, by equation ($\star$), this means that $card([\langle e,0\rangle]_S) \neq e+1$.

On the other hand, suppose that R_e has no equivalence classes of size $e+1$. This can happen in two cases that shall be considered separately.

First, assume that there is a least stage s_0 such that any $R_e[s]$, with $s \geq s_0$, has no equivalence classes of size $e+1$. If so, by reasoning as above, we obtain that there is a least stage $\langle e,t\rangle$ after which we have $e+1$ many e-witnesses and, in all further stages focusing on P_e, we enter in case (4). Therefore, $card([\langle e,0\rangle]) = card(Y_e) = e+1$, while R_e has no equivalence classes of size $e+1$.

Secondly, assume that there are infinitely many stages at which $R_e[s]$ has an equivalence class of size $e+1$, even though there is no R_e-class of such size. This can happen only if the approximation to R_e has infinitely many mindchanges regarding the oldest equivalence class of size $e+1$. That is, only if P_e turns **on** infinitely often, and therefore the construction enters in case (3) infinitely often when dealing with P_e. We claim that this makes $Y_e = Y_e[0]$, and therefore $card(Y_e) = e+1$. To see this, suppose that there is a least z such that $z \in Y_e \smallsetminus Y_e[0]$. This means that there is stage s_0 such that, for all $s \geq s_0$, $z \in Y_e[s]$. Yet, it is not hard to see that there must be a stage $\langle e, n \rangle \geq s_0$ in which P_e is **on** and u_e takes value z. Hence, in this stage, the construction enters in case (3) which implies that $z \notin Y_e[s+1]$, a contradiction. So, $card(Y_e) = e + 1$. By equation $(\star)$, it follows that $card([\langle e, 0 \rangle]_S) = e + 1$, while R_e has no equivalence class of such size. □

This concludes the proof of Theorem 2.1. □

Having shown that there is a Π^0_1 equivalence structure which has no Σ^0_1 copy, it comes natural to investigate the relation between the Π^0_1 and the Δ^0_2 equivalence structures. Obviously, every Π^0_1 (or Σ^0_1) equivalence structure is isomorphic to a Δ^0_2 equivalence structure. One might conjecture that the inclusion is strict, i.e., that there is a Δ^0_2 equivalence structure which is not isomorphic to any Π^0_1 equivalence structure. This is not the case.

Theorem 2.2. *The isomorphism types of* Π^0_1 *and* Δ^0_2 *equivalence structures coincide. That is,* $\mathrm{IsoType}(\mathbb{EQ}, \Pi^0_1) = \mathrm{IsoType}(\mathbb{EQ}, \Delta^0_2)$.

Proof. Let E be a Δ^0_2 equivalence relation. If E has only finitely many equivalence classes, then it is obvious that E is isomorphic to a computable structure. Hence, without loss of generality, we assume that E has infinitely many classes.

A relativized version of Theorem 1.1 implies that there is a $\mathbf{0}'$-limitwise monotonic function $H\colon \omega \to \omega \cup \{\infty\}$ with the following property: For any $\alpha \in \omega \cup \{\infty\}$, the number of E-equivalence classes of size α is equal to

$$\gamma(\alpha) := card(\{k \in \omega : H(k) = \alpha\}).$$

Note that $H(k)$ cannot be equal to zero.

By Theorem 1.2, one can choose a computable function $g\colon \omega^2 \to \omega$ such that for any $k \in \omega$, we have

$$H(k) = \liminf_s g(k, s).$$

Since $H(k) \neq 0$, we may assume that $g(k, s) \geq 1$ for all s (i.e., if $g(k, s)$ equals zero, then one just replaces this zero with one).

We give a construction of a Π^0_1 equivalence relation R. As per usual, we define R by enumerating its complement, i.e., at any stage of the construction, we are allowed to declare that $(x \not R\, y)$, where $x \neq y$. Surely, if $(x \not R\, y)$, then we always (implicitly) assume $(y \not R\, x)$. In other words, we can only provide negative information about R, and we never give positive pieces of R-data.

At a non-zero stage s, we define a finite set $A[s]$ and a function $\ell[s]$ acting from $A[s]$ onto $\{0, 1, \ldots, s-1\}$. The intuition behind $\ell[s]$ is as follows—at the stage s, we think that the first s classes of $R[s]$ are precisely the following sets:

$$\{x \in A[s] : \ell(x)[s] = k\}, \; k = 0, 1, \ldots, s-1.$$

Thus, we will assume that we "automatically" declare the following R-information: if $\ell(x)[s]\downarrow \neq \ell(y)[s]\downarrow$, then $(x \not{R}\, y)$.

At a stage s, an element x is called *fresh* if $x \notin \bigcup_{t<s} A[s]$.

The construction

Stage 0. Set $A[0] := \emptyset$.

Stage $s+1$. First, choose an element w as follows:

- If the set $\{x \text{ is non-fresh} : x \notin A[s]\}$ is non-empty, then w is the least element of this set.
- Otherwise, w is the least fresh element.

Declare $\ell(w)[s+1] := s$.

After that, for each $k < s$, we act according to the following instructions. We define

$$\delta(k) := card(\{x \in A[s] : \ell(x)[s] = k\}).$$

(1) If $g(k, s+1) > \delta(k)$, then choose the least $(g(k, s+1) - \delta(k))$ fresh elements. For each of these elements y, set $\ell(y)[s+1] := k$.
(2) If $g(k, s+1) < \delta(k)$, then find the greatest $(\delta(k) - g(k, s+1))$ elements z such that $\ell(z)[s] = k$. For each of these z, set $\ell(z)[s+1]$ undefined. Surely, we also assume $z \notin A[s+1]$.

The description of the construction is finished.

The verification

It is clear that the constructed relation R is co-c.e. Moreover, it is not hard to verify the following claim (recall that $g(k, s) \geq 1$ for all k and s).

Claim 2.1. *Every $x \in \omega$ satisfies precisely one of the following two cases:*

(a) There is a stage s_0 such that $\ell(x)[s_0]$ is defined, and

$$\ell(x)[s] = \begin{cases} \text{undefined}, & \text{if } s < s_0, \\ l(x)[s_0], & \text{if } s \geq s_0. \end{cases}$$

(b) There are stages $s_0 < s_1 < s_2$ *such that* $\ell(x)[s_0]\downarrow < \ell(x)[s_2]\downarrow$, *and*

$$\ell(x)[s] = \begin{cases} \textit{undefined}, & \textit{if } s < s_0, \\ l(x)[s_0], & \textit{if } s_0 \leq s < s_1, \\ \textit{undefined}, & \textit{if } s_1 \leq s < s_2, \\ l(x)[s_2], & \textit{if } s \geq s_2. \end{cases}$$

For an element $x \in \omega$, let $\ell^*(x) := \lim_s \ell(x)[s]$. It is not difficult to show that the following conditions are equivalent:

$$(x \, R \, y) \; \Leftrightarrow \; \ell^*(x) = \ell^*(y). \tag{1}$$

Therefore, we deduce that R is an equivalence relation.

At the end of a stage $s+1$, the relation $R[s+1]$ satisfies the following: for any $k < s$,

$$card(\{x \in A[s+1] : \ell(x)[s+1] = k\}) = g(k, s+1).$$

This fact and the description of the construction together imply that for any $k \in \omega$ and any $m \geq 1$, the following are equivalent:

(1) $card(\{x \in \omega : \ell^*(x) = k\}) \geq m$;
(2) $\exists s_0 (\forall s \geq s_0)(g(k,s) \geq m)$.

Thus, we obtain that

$$card(\{x \in \omega : \ell^*(x) = k\}) = \liminf_s \, g(k,s) = H(k).$$

By (1), for each $\alpha \in \omega \cup \{\infty\}$, the relations R and E have the same number of equivalence classes of size α. Therefore, R is isomorphic to E. Theorem 2.2 is proved. □

It is not hard to see that the last two theorems relativize, and the same is true for the proof of Cenzer, Harizanov, and Remmel[8] that there is a Σ^0_1 equivalence structure with no computable copy. Hence, we have the following characterization of the isomorphism types of equivalence structures realized at various levels of the arithmetical hierarchy.

Theorem 2.3. *For all* $n > 0$, *we have that*

$$\mathrm{IsoType}(\mathbb{EQ}, \Sigma^0_n) \subsetneq \mathrm{IsoType}(\mathbb{EQ}, \Pi^0_n) = \mathrm{IsoType}(\mathbb{EQ}, \Delta^0_{n+1}).$$

We conclude our study of equivalence structures by lifting our focus to the analytical hierarchy. We show that for all $n > 0$,

$$\mathrm{IsoType}(\mathbb{EQ}, \Delta^1_n) \subsetneq \mathrm{IsoType}(\mathbb{EQ}, \Sigma^1_n);$$
$$\mathrm{IsoType}(\mathbb{EQ}, \Delta^1_n) \subsetneq \mathrm{IsoType}(\mathbb{EQ}, \Pi^1_n);$$
$$\mathrm{IsoType}(\mathbb{EQ}, \Sigma^1_n) \cup \mathrm{IsoType}(\mathbb{EQ}, \Pi^1_n) \subsetneq \mathrm{IsoType}(\mathbb{EQ}, \Delta^1_{n+1}).$$

Proposition 2.1. *Let n be a non-zero natural number.*

(i) There is a Σ^1_n equivalence structure which is not isomorphic to any Δ^1_n equivalence structure. A similar result holds for "Π^1_n vs. Δ^1_n".
(ii) There is a Δ^1_{n+1} equivalence structure such that its isomorphism type cannot be realized neither by a Σ^1_n structure, nor by a Π^1_n structure.

Proof. *Ad (i).* Consider a computable partition of ω into infinitely many finite blocks: for an index $i \in \omega$, the ith block contains precisely $2i + 4$ elements — $a^i_0, a^i_1, \ldots, a^i_{2i+3}$.

Let $X \subseteq \omega$ be a set belonging to $\Sigma^1_n \setminus \Delta^1_n$. We define an equivalence structure S as follows:

(a) For every $i \in \omega$, the elements $a^i_0, a^i_1, \ldots, a^i_{2i+2}$ are S-equivalent. If $i \neq j$, then a^i_k and a^j_ℓ are not S-equivalent.
(b) $a^i_{2i+3} \in [a^i_0]_S$ if and only if $i \in X$.

A standard application of the Tarski–Kuratowski algorithm shows that the relation S is Σ^1_n.

Note that the character $\chi(S)$ satisfies the following:

$$\chi(S) = \{(1, k) : k \geq 1\} \cup \{(2i + 4, 1) : i \in X\} \cup \{(2j + 3, 1) : j \notin X\}.$$

Thus, $X \leq_T \chi(S)$ and the set $\chi(S)$ cannot be Δ^1_n.

On the other hand, for an arbitrary countable equivalence structure R, the character $\chi(R)$ is Σ^0_2 in the atomic diagram of R. Hence, if R is a Δ^1_n structure, then $\chi(R) \in \Delta^1_n$. Therefore, we deduce that our structure S is not isomorphic to any Δ^1_n equivalence structure.

The proof of the case "Π^1_n vs. Δ^1_n" is essentially the same: just choose X belonging to $\Pi^1_n \setminus \Delta^1_n$.

Ad (ii). Let A be an m-complete Σ^1_n set. We define $X := A^{(3)}$, and we build an equivalence structure S by applying the previous construction to the set X. Since $A \in \Delta^1_{n+1}$, we deduce that both X and S are also Δ^1_{n+1}.

For proof by contradiction, assume that R is a Σ^1_n equivalence structure, which is isomorphic to S. By $D(R)$ we denote the atomic diagram of R. It is clear that $D(R)$ is computable in the Σ^1_n-complete set A. Therefore,

$$A^{(3)} \leq_T \chi(S) = \chi(R) \leq_T D(R)^{(2)} \leq_T A^{(2)},$$

which gives a contradiction. Hence, the isomorphism type of S cannot be realized by a Σ^1_n structure. Moreover, a similar argument applies for Π^1_n structures. □

3. Isomorphism types of preorders

This section is devoted to the case of preorders. First, observe that there is a Π^0_1 preorder that is not isomorphic to any Σ^0_1 preorder: this is an immediate consequence of Theorem 2.1. The next theorem says that the converse also holds, thus distinguishing the case of preorders from that of equivalence structures.

Theorem 3.1. *There is a Σ^0_1 preorder R, which is not isomorphic to any Π^0_1 preorder.*

Proof. First, we describe how one can construct a sufficiently large class of Σ^0_1 preorders. After that, inside this class, we will choose a preorder satisfying the theorem.

Let $B \subseteq \omega$ be an arbitrary Δ^0_2 set with $0 \notin B$. Choose a total computable function $g(x, s)$ such that $B(x) = \lim_s g(x, s)$.

We define a Σ^0_1 preorder S_B as follows. Fix a computable partition of ω into

$$\{c, d\} \cup \{a_i : i \in \omega\} \cup \{b_j : j \in \omega\}.$$

By $x <_{S_B} y$ we denote the formula $(x\ S_B\ y)\&(y \not S_B\ x)$.

Beforehand, we set:

- The relation S_B is reflexive.
- All a_i, $i \in \omega$, are pairwise S_B-incomparable. For every i, $c <_{S_B} a_i$, and a_i is S_B-incomparable with d.
- The element d is S_B-incomparable with c. For every $j \in \omega$, we have $b_{j+1} <_{S_B} b_j <_{S_B} d$, and b_j is incomparable with c.

In order to finish the definition of S_B, our construction will define the following finite values: for every $i \in \omega$, one needs to recover

$$v(i) := \min\{j \in \omega : b_j <_{S_B} a_i\}.$$

The Σ^0_1-ness of the preorder S_B will be ensured by the step-by-step definition of $v(i)$, which proceeds as follows:

- First, set $v(i)[0]$ as undefined.
- Then at some stage s_0, we put $v(i)[s_0] := k_0$, i.e., we enumerate the formulas saying

$$(b_{k_0}\ S_B\ a_i),\ (b_{k_0+1}\ S_B\ a_i),\ (b_{k_0+2}\ S_B\ a_i),\ \dots$$

  into our (approximation of) S_B.
- After that, the value $v(i)[s]$ can change at most one time: at a stage $s_1 > s_0$, we put $v(i)[s_1] := 0$ — this means that we enumerate the facts $(b_0\ S_B\ a_i)$, $(b_1\ S_B\ a_i)$, $\dots$, $(b_{k_0-1}\ S_B\ a_i)$ into S_B.

At a stage $s+1$, we say that an element a_i is *fresh* if the value $v(i)[s]$ is undefined.

The construction

Stage 0. Set $v(i)[0]$ as undefined for all $i \in \omega$.

Stage $s+1 = 2t+1$. Find the least fresh a_i, and define $v(i)[s+1] := 0$.

Stage $s+1 = 2t+2$. For each non-zero $x \leq s$, proceed as follows:

- If $g(x,s) = 0$ and there is a_i with $v(i)[s] = x$, then set $v(i)[s+1] := 0$ (for all such a_i).
- If $g(x,s) = 1$ and there is no a_i with $v(i)[s] = x$, then choose the least fresh a_k and define $v(k)[s+1] := x$.

The verification

It is clear that for every $i \in \omega$, there exists the limit $v(i) := \lim_s v(i)[s]$. Moreover, consider the first stage s_0 such that $v(i)[s_0]$ is defined. Then precisely one of the following two cases holds:

(i) either for all $s \geq s_0$, we have $v(i)[s] = v(i)[s_0] = v(i)$, or
(ii) $v(i)[s_0] \neq 0$ and there is a stage $s_1 > s_0$ such that

$$v(i)[s] = \begin{cases} v(i)[s_0], & \text{if } s_0 \leq s < s_1, \\ 0, & \text{if } s \geq s_1. \end{cases}$$

Therefore, one can deduce that the preorder S_B is c.e.

The fact above and the description of the construction together imply the following:

Claim 3.1. *Let x be a non-zero natural number.*

- *If $x \in B$, then there is a unique i with $v(i) = x$.*

- *If $x \notin B$, then $v(i) \neq x$ for all $i \in \omega$.*

Moreover, there are infinitely many k with $v(k) = 0$.

This claim shows that the isomorphism type of the preorder S_B depends only on the set B, but not on the choice of its approximation g.

By Theorem 1.3, there is a d.c.e. set A that is not limitwise monotonic. Without loss of generality, we may assume that $0 \notin A$. We prove that the preorder $R := S_A$ is not isomorphic to any Π^0_1 preorder.

Towards a contradiction, assume that Q is a Π^0_1 preorder, which is isomorphic to R. Fix an isomorphism $f\colon R \cong Q$. We will slightly abuse our notations, and identify the elements c and $f(c)$, b_i and $f(b_i)$, etc.

In a non-uniform way, one can find the elements c and d inside Q. We note that:

- If $w \in \{a_i : i \in \omega\}$ inside Q, then w is Q-incomparable with d.
- If $w \in \{b_j : j \in \omega\}$, then w is Q-incomparable with c.
- Suppose that u and w are different elements from $\{b_j : j \in \omega\}$. Then u, w satisfy precisely one of the following: either $u <_Q w$, or $w <_Q u$.

Since the preorder Q is co-c.e., we deduce the following: one can effectively check whether a given w is an f-image of some a_i or of some b_j. Moreover, the ordering of b_j-s inside Q is recursive.

Consider a computable set $U := \{f(a_i) : i \in \omega\}$, and for each $w \in U$, define the value

$$p(w) := \text{the number of } b_j\text{-s, which are } Q\text{-incomparable with } w.$$

Since Q is Π^0_1, the function p is limitwise monotonic. Hence, the set

$$C := \{p(w) : w \in U,\ p(w) \geq 1\}$$

is also limitwise monotonic.

On the other hand, recall that Q is isomorphic to $R = S_A$, and hence, by Claim 3.1, C must be equal to A, which contradicts the choice of A. Therefore, the preorder R is not isomorphic to a Π^0_1 preorder. □

Acknowledgments

The work of Bazhenov was carried out within the framework of the state contract of the Sobolev Institute of Mathematics (project no. FWNF-2022-0011). San Mauro was supported by the Austrian Science Fund FWF, project M 2461.

References

1. A. T. Nurtazin, Strong and weak constructivization and computable families, *Algebra and Logic* **13**, 177 (1974).
2. S. Tennenbaum, Non-archimedean models for arithmetic, *Notices of the American Mathematical Society* **6**, p. 44 (1959).
3. N. G. Khisamiev, Constructive abelian groups, in *Studies in Logic and the Foundations of Mathematics*, (Elsevier, 1998) pp. 1177–1231.
4. C. Spector, Recursive well-orderings, *The Journal of Symbolic Logic* **20**, 151 (1955).
5. W. Calvert, D. Cenzer, V. Harizanov and A. Morozov, Effective categoricity of equivalence structures, *Annals of Pure and Applied Logic* **141**, 61 (2006).
6. R. G. Downey, A. G. Melnikov and K. M. Ng, A Friedberg enumeration of equivalence structures, *Journal of Mathematical Logic* **17**, p. 1750008 (2017).
7. S. S. Goncharov and J. F. Knight, Computable structure and non-structure theorems, *Algebra and Logic* **41**, 351 (2002).
8. D. Cenzer, V. Harizanov and J. B. Remmel, Σ_1^0 and Π_1^0 equivalence structures, *Annals of Pure and Applied Logic* **162**, 490 (2011).
9. S. Gao and P. Gerdes, Computably enumerable equivalence relations, *Studia Logica* **67**, 27 (2001).
10. U. Andrews, S. Lempp, J. S. Miller, K. M. Ng, L. San Mauro and A. Sorbi, Universal computably enumerable equivalence relations, *The Journal of Symbolic Logic* **79**, 60 (2014).
11. S. Coskey, J. D. Hamkins and R. Miller, The hierarchy of equivalence relations on the natural numbers under computable reducibility, *Computability* **1**, 15 (2012).
12. U. Andrews and A. Sorbi, Joins and meets in the structure of ceers, *Computability* **8**, 193 (2019).
13. E. Ianovski, R. Miller, K. M. Ng and A. Nies, Complexity of equivalence relations and preorders from computability theory, *The Journal of Symbolic Logic* **79**, 859 (2014).
14. E. Fokina, D. Rossegger and L. San Mauro, Measuring the complexity of reductions between equivalence relations, *Computability* **8**, 265 (2019).
15. E. B. Fokina, S.-D. Friedman, V. Harizanov, J. F. Knight, C. McCoy and A. Montalbán, Isomorphism relations on computable structures, *The Journal of Symbolic Logic* **77**, 122 (2012).
16. K. M. Ng and H. Yu, On the degree structure of equivalence relations under computable reducibility, *Notre Dame Journal of Formal Logic* **60**, 733 (2019).
17. N. Bazhenov, M. Mustafa, L. San Mauro, A. Sorbi and M. Yamaleev, Classifying equivalence relations in the Ershov hierarchy, *Archive for Mathematical Logic* **59**, 835 (2020).
18. F. Montagna and A. Sorbi, Universal recursion theoretic properties of r.e. preordered structures, *The Journal of Symbolic Logic* **50**, 397 (1985).
19. U. Andrews and A. Sorbi, Effective inseparability, lattices, and preordering relations, *The Review of Symbolic Logic* **14**, 838 (2021).
20. G. Metakides and A. Nerode, Recursively enumerable vector spaces, *Annals*

of Mathematical Logic **11**, 147 (1977).
21. V. Selivanov, Positive structures, in *Computability and Models*, (Springer, 2003) pp. 321–350.
22. A. Gavruskin, S. Jain, B. Khoussainov and F. Stephan, Graphs realised by r.e. equivalence relations, *Annals of Pure and Applied Logic* **165**, 1263 (2014).
23. E. Fokina, B. Khoussainov, P. Semukhin and D. Turetsky, Linear orders realized by c.e. equivalence relations, *The Journal of Symbolic Logic* **81**, 463 (2016).
24. L. Feiner, Orderings and Boolean algebras not isomorphic to recursive ones (1967), MIT. Ph.D. Thesis.
25. R. G. Downey, A. M. Kach and D. Turetsky, Limitwise monotonic functions and their applications, in *Proceedings of the 11th Asian Logic Conference*, eds. T. Arai, Q. Feng, B. Kim, G. Wu and Y. Yang (World Scientific, Singapore, 2011) pp. 59–85.
26. K. Harris, η-representation of sets and degrees, *J. Symb. Log.* **73**, 1097 (2008).
27. A. M. Kach, Computable shuffle sums of ordinals, *Arch. Math. Logic* **47**, 211 (2008).
28. B. Khoussainov, A. Nies and R. A. Shore, Computable models of theories with few models, *Notre Dame J. Formal Logic* **38**, 165 (1997).

Rogers semilattices in the analytical hierarchy: The case of finite families

Nikolay Bazhenov

Sobolev Institute of Mathematics,
4 Acad. Koptyug Avenue,
Novosibirsk, 630090, Russia;
Novosibirsk State University,
2 Pirogova Street,
Novosibirsk, 630090, Russia
E-mail: bazhenov@math.nsc.ru

Manat Mustafa

Department of Mathematics,
School of Science and Humanities,
Nazarbayev University,
53 Qabanbay Batyr Avenue,
Astana, 010000, Kazakhstan
E-mail: manat.mustafa@nu.edu.kz

A numbering of a countable family S is a surjective map from the set of natural numbers ω onto S. This paper studies Rogers semilattices, i.e. upper semilattices induced by the reducibility between numberings, for families $S \subset P(\omega)$. Working in set theory **ZF+DC+PD**, we obtain the following results on families from various levels of the analytical hierarchy.

For a non-zero number n, by E^1_n we denote Π^1_n if n is odd, and Σ^1_n if n is even. We show that for a finite family S of E^1_n sets, its Rogers E^1_n-semilattice has the greatest element if and only if S contains the least element under set-theoretic inclusion. Furthermore, if S does not have the $\subseteq$-least element, then the corresponding Rogers E^1_n-semilattice is upwards dense.

Keywords: Theory of numberings, analytical hierarchy, upper semilattice, Rogers semilattice, universal numbering, minimal cover, elementary theory, projective determinacy, axiom of constructibility.

1. Introduction

Let $\mathcal{S}$ be a countable set. A *numbering* of $\mathcal{S}$ is a surjective map ν from the set of natural numbers ω onto $\mathcal{S}$. The origins of the theory of numberings can be traced back to the works of Gödel[23] and Kleene[29]. The proof of Gödel's incompleteness theorems uses an effective numbering of first-order formulae. Kleene (see Theorem XXII in § 65 of Ref.[29]) gave a construction of a universal partial computable function — this result provides a universal computable numbering for the family of all unary partial computable functions. At the end of 1950s, the foundations of the modern theory of numberings were developed by Kolmogorov and Uspenskii[31,42] and, independently, by Rogers[37].

The algorithmic complexity of different numberings is typically compared via the notion of *reducibility* between numberings: A numbering ν is *reducible* to a numbering μ, denoted by $\nu \leq \mu$, if there is total computable function $f(x)$ such that $\nu(n) = \mu(f(n))$ for all $n \in \omega$. More informally, there is an effective procedure which, given a ν-index of an object from $\mathcal{S}$, computes a μ-index for the same object.

Since the end of 1960s, the research in the theory of numberings has been mainly focused in the area of *Rogers semilattices.* We give a very brief overview of the classical setting in this area. From now on, we consider only families $\mathcal{S}$ containing subsets of ω, i.e., we always assume that $\mathcal{S} \subset P(\omega)$ and $\mathcal{S}$ is countable.

Let $\mathcal{S}$ be a family of computably enumerable (c.e.) sets. A numbering ν of the family $\mathcal{S}$ is *computable* if the set

$$G_\nu = \{\langle n, x\rangle : x \in \nu(n)\} \tag{1}$$

is c.e. A family $\mathcal{S}$ is *computable* if it has a computable numbering. In other words, the computability of $\mathcal{S}$ means that one can *uniformly enumerate* all sets from $\mathcal{S}$. Note that in general, this enumeration *allows* repetitions.

As simple examples of computable families, one can immediately recall the family of all finite sets and the family of all c.e. sets. A more delicate example can be constructed as follows. If a number $e \notin \emptyset'$, then our family $\mathcal{T}$ contains one-element sets $\{2e\}$ and $\{2e+1\}$. If $e \in \emptyset'$, then the set $\{2e, 2e+1\}$ belongs to $\mathcal{T}$. It is easy to see that the constructed family $\mathcal{T}$ admits a uniform enumeration and thus, $\mathcal{T}$ is computable. A more interesting feature of $\mathcal{T}$ is the following: *Any* computable numbering ν of $\mathcal{T}$ must have repetitions, i.e. there are indices $m \neq n$ with $\nu(m) = \nu(n)$.

In a standard recursion-theoretic way, the notion of reducibility between numberings gives rise to the corresponding upper semilattice: For a computable family $\mathcal{S}$, the *Rogers semilattice* of $\mathcal{S}$ contains the degrees of all computable numberings of $\mathcal{S}$. As usual, here two numberings have the same degree if they are reducible to each other. Roughly speaking, the supremum of two numberings is provided by their join, see Sec. 2.1 for formal details.

To give a flavor of studies of computable families, we mention here two celebrated classical results on Rogers semilattices: Let $\mathcal{S}$ be a computable family, and let R be its Rogers semilattice. Khutoretskii[28] proved that if R contains more than one element, then R is infinite. Selivanov[40] showed that an infinite R cannot be a lattice.

Goncharov and Sorbi[24] started developing the theory of *generalized computable* numberings. One of their approaches to generalized computations can be summarized as follows. Let Γ be a complexity class (e.g., Σ^0_1, d-Σ^0_1, Σ^0_n, or Π^1_n). A numbering ν of a family $\mathcal{S}$ is Γ-*computable* if the set G_ν from Eq. (1) belongs to the class Γ. We say that a family $\mathcal{S}$ is Γ-*computable* if it has a Γ-computable numbering. Note that the classical notion of a computable numbering becomes a synonym for a Σ^0_1-computable numbering.

In a similar way to computable families, one can define the Rogers semilattice $\mathcal{R}_\Gamma(\mathcal{S})$ for a Γ-computable family $\mathcal{S}$, see Sec. 2.1 for formal details.

We follow the approach of Goncharov and Sorbi, and study the following problem:

Problem 1.1. *Let Γ be a class of the analytical hierarchy, i.e. $\Gamma \in \{\Sigma^1_n, \Pi^1_n : n \geq 1\}$. Study the elementary theories of Rogers semilattices for Γ-computable families.*

The current paper is a continuation of studies developed in Refs.[12,14]. These papers concentrated on Rogers semilattices of Π^1_n-computable families. Dorzhieva[14] showed that for a Π^1_n-computable family $\mathcal{S}$, one of the following two conditions holds:

(a) either the Rogers semilattice $\mathcal{R}_{\Pi^1_n}(\mathcal{S})$ contains only one element, or
(b) the first-order theory $Th(\mathcal{R}_{\Pi^1_n}(\mathcal{S}))$ is hereditarily undecidable.

The article[12] proves the following: if the semilattice $\mathcal{R}_{\Pi^1_n}(\mathcal{S})$ contains more than one element, then for any non-zero $m \neq n$ and any Π^1_m-computable family $\mathcal{T}$, the structure $\mathcal{R}_{\Pi^1_n}(\mathcal{S})$ is not isomorphic to $\mathcal{R}_{\Pi^1_m}(\mathcal{T})$. Further related work is discussed in Sec. 2.2.

The paper[12] left the following problem open:

Problem 1.2. *Let n be a non-zero natural number. Consider Rogers semilattices $\mathcal{R}_{\Pi^1_n}(\mathcal{S})$ for Π^1_n-computable families $\mathcal{S}$. How many isomorphism types do these semilattices realize?*

While attacking Problem 1.2, we observed the following: In order to apply the known numbering-theoretic techniques in this setting, one *needs* to employ additional set-theoretic assumptions.

This observation motivated us to organize our paper as follows: We prove a number of results concerning Problem 1.1 under the assumption of *Projective Determinacy* (**PD**, see Sec. 3.2 for the details). Why did we choose **PD** as an additional axiom? Tanaka[41] already initiated a systematic development of recursion theory for the levels of analytical hierarchy, under the assumption of **PD**. We found his approach well-suited to our goals.

In order to make our paper accessible to both numbering-theoretic and set-theoretic communities, we tried to make the exposition as self-contained as possible.

The structure of the paper is as follows. Sec. 2 contains the necessary preliminaries on the theory of numberings. Sec. 3 discusses the background on the analytical hierarchy. In particular, Sec. 3.2 introduces consequences of **PD**, which will be employed in our proofs. It is well-known that under **PD**, the levels of analytical hierarchy exhibit "flip-flopping" behavior: kindred levels are Π^1_1, Σ^1_2, Π^1_3, $\Sigma^1_4, \ldots$ (see, e.g., Refs.[25,34]). Thus, following Refs.[2,41], we use the following conventions:

- For a number $k \in \omega$, $E^1_{2k+1} := \Pi^1_{2k+1}$ and $E^1_{2k+2} := \Sigma^1_{2k+2}$.
- For a non-zero $n \in \omega$, we consider E^1_n-computable families $\mathcal{S}$. By $\mathcal{R}^1_n(\mathcal{S})$ we denote the Rogers semilattice $\mathcal{R}_{E^1_n}(\mathcal{S})$.

Sec. 4 studies elementary properties of the semilattices $\mathcal{R}^1_n(\mathcal{S})$ for *finite* families $\mathcal{S}$. Any such $\mathcal{R}^1_n(\mathcal{S})$ is a distributive upper semilattice (Proposition 4.1). Note that the proof of this fact does not require **PD**. While assuming **PD**, we obtain a criterion for when $\mathcal{R}^1_n(\mathcal{S})$ has the greatest element (Theorem 4.1). We also show that if $\mathcal{R}^1_n(\mathcal{S})$ has no greatest element, then it is upwards dense (Theorem 4.2).

Sec. 5 discusses first-order properties of $\mathcal{R}^1_n(\mathcal{S})$ for *infinite* families $\mathcal{S}$. We make use of Dorzhieva's results from Ref.[14] to establish the following:

(1) For an infinite E^1_n-computable family $\mathcal{S}$, the upper semilattice $\mathcal{R}^1_n(\mathcal{S})$ is not weakly distributive (Theorem 5.1). Consequently, $\mathcal{R}^1_n(\mathcal{S})$ is not distributive.
(2) For an *arbitrary* E^1_n-computable family $\mathcal{S}$ which contains at least two elements, the semilattice $\mathcal{R}^1_n(\mathcal{S})$ is infinite, and it is not a lattice (Corollary 5.1).

Summarizing, our results (together with Lemma 3.1 below) provide a first step to the solution of Problem 1.2 — now we know that under **PD**, there are at least four different isomorphism types of Rogers E^1_n-semilattices $\mathcal{R}^1_n(\mathcal{S})$, induced via the following families:

(1) A one-element family $\mathcal{S}$ — in this case, the structure $\mathcal{R}^1_n(\mathcal{S})$ is also one-element.
(2) A finite family $\mathcal{S}$, containing more than one element and possessing the $\subseteq$-least element.
(3) A finite family $\mathcal{S}$ without the $\subseteq$-least element.
(4) An infinite E^1_n-computable family $\mathcal{S}$.

The last section discusses further problems. In particular, we consider the following question: What happens to our results, if we replace **PD** with the Axiom of Constructibility ($V = L$)?

Recall that the *Axiom of Dependent Choices* (**DC**) states the following: For any non-empty set A and any set of pairs $P \subseteq A \times A$, we have:

$$(\forall x \in A)(\exists y \in A)P(x,y) \;\Rightarrow\; (\exists f\colon \omega \to A)(\forall n)P(f(n), f(n+1)).$$

Throughout the paper, we work in set theory **ZF**+**DC**.

2. Preliminaries

Lower-case letters $x, y, z, \dots$ denote variables that range over ω. Capital letters $X, Y, Z, \dots$ are used for subsets of ω.

By $\leq_\omega$ we denote the standard ordering of natural numbers. Recall that ω^ω is the set of all total functions acting from ω to ω.

As usual, $\langle \cdot, \cdot \rangle$ is a standard pairing function over ω. By $\langle \cdot \rangle_0$ and $\langle \cdot \rangle_1$ we denote computable functions such that for every $n \in \omega$, we have $\langle \langle n \rangle_0, \langle n \rangle_1 \rangle = n$.

We treat upper semilattices as structures in the language $L_{usl} = \{\leq, \vee\}$.

2.1. *Numberings*

Suppose that ν is a numbering of a family $\mathcal{S}_0$, and μ is a numbering of a family $\mathcal{S}_1$. Notice that the condition $\nu \leq \mu$ always implies that $\mathcal{S}_0 \subseteq \mathcal{S}_1$.

Numberings ν and μ are *equivalent*, denoted by $\nu \equiv \mu$, if $\nu \leq \mu$ and $\mu \leq \nu$. The numbering $\nu \oplus \mu$ of the family $\mathcal{S}_0 \cup \mathcal{S}_1$ is defined as follows:

$$(\nu \oplus \mu)(2x) = \nu(x), \quad (\nu \oplus \mu)(2x+1) = \mu(x).$$

The following fact is well-known (see, e.g., p. 36 in Ref.[16]): If ξ is a numbering of a family $\mathcal{T}$, then

$$(\nu \leq \xi \,\&\, \mu \leq \xi) \;\Leftrightarrow\; (\nu \oplus \mu \leq \xi).$$

For further background on numberings, the reader is referred to, e.g., Refs.[7,16,17,20–22].

Let Γ be a complexity class with the following properties:

(a) If ν is a Γ-computable numbering and μ is a numbering such that $\mu \leq \nu$, then μ is Γ-computable.
(b) If numberings ν and μ are both Γ-computable, then the numbering $\nu \oplus \mu$ is also Γ-computable.

Note that it is not hard to show that for a non-zero natural number n, each of the classes Σ^0_n, Σ^{-1}_n, and Π^1_n has these properties.

Let $\mathcal{S}$ be a Γ-computable family. By $Com_\Gamma(\mathcal{S})$ we denote the set of all Γ-computable numberings of $\mathcal{S}$. Since the relation $\equiv$ is a congruence on the structure $(Com_\Gamma(\mathcal{S}); \leq, \oplus)$, we use the same symbols $\leq$ and $\oplus$ on numberings and on their $\equiv$-equivalence classes.

The quotient structure $\mathcal{R}_\Gamma(\mathcal{S}) := (Com_\Gamma(\mathcal{S})/{\equiv}; \leq, \oplus)$ is an upper semilattice. We say that $\mathcal{R}_\Gamma(\mathcal{S})$ is the *Rogers semilattice* of the Γ-computable family $\mathcal{S}$.

2.2. *Related work*

There is a large body of literature which studies a counterpart of Problem 1.1 in the setting of the arithmetical hierarchy. For the sake of brevity, here we use the term *Rogers* Σ^0_n*-semilattice* as a synonym for "the Rogers semilattice of a Σ^0_n-computable family."

Ershov and Lavrov[19] (see also p. 72 in Ref.[16], and Ref.[18]) showed that there are finite families $\mathcal{S}_i$, $i \in \omega$, of c.e. sets such that the semilattices $\mathcal{R}_{\Sigma^0_1}(\mathcal{S}_i)$ are pairwise non-isomorphic. In other words, there are infinitely many isomorphism types of Rogers Σ^0_1-semilattices. V'yugin[43] proved that there are infinitely many pairwise elementarily non-equivalent Rogers Σ^0_1-semilattices. Badaev, Goncharov, and Sorbi[8] proved that for any natural number $n \geq 2$, there are infinitely many pairwise elementarily non-equivalent Rogers Σ^0_n-semilattices. The reader is referred to, e.g., Refs.[3,4,9,11,36] for further results on Rogers Σ^0_n-semilattices.

Recall that a numbering ν is *Friedberg* if $\nu(k) \neq \nu(m)$ for all $k \neq m$. Dorzhieva[13,15] studied Friedberg numberings for families of sets in the analytical hierarchy.

Kalimullin, Puzarenko, and Faizrakhmanov[26,27] considered *computable* Π^1_1*-numberings*. A Π^1_1*-numbering* of a family $\mathcal{S}$ is a *partial* map ν acting from ω onto $\mathcal{S}$ such that the domain of ν is enumeration reducible to the

Π^1_1-complete set $\mathcal{O}$. A Π^1_1-numbering ν is *computable* if the set

$$G^*_\nu = \{\langle n, x\rangle : n \in dom(\nu),\ x \in \nu(n)\}$$

is enumeration reducible to $\mathcal{O}$.

3. Background on the analytical hierarchy

Here we discuss known results on the analytical hierarchy, which will be employed in our proofs. Furthermore, the section includes proofs of several useful (but a little bit technical) results on numberings: Lemma 3.1, Proposition 3.1, and Lemma 3.2.

Recall that a predicate $R(x_1, \dots, x_m; f_1, \dots, f_n)$, where the variables f_i denote elements from ω^ω, is *recursive* if there is an index $e \in \omega$ such that for all $f_1, \dots, f_n$ and $x_1, \dots, x_m$, the following conditions hold:

(a) the value $\varphi_e^{f_1 \oplus \dots \oplus f_n}(x_1, \dots, x_m)$ is defined;
(b) the predicate $R(x_1, \dots, x_m; f_1, \dots, f_n)$ is true if and only if $\varphi_e^{f_1 \oplus \dots \oplus f_n}(x_1, \dots, x_m) = 1$.

If an index e satisfies these conditions, then we say that e *witnesses the recursiveness* of the predicate R.

We follow Ref.[39] and use the following version of a *normal form* for analytical subsets of ω: Let n be a non-zero natural number. A set $X \subseteq \omega^m$ is Π^1_n if and only if there is a recursive predicate $R(x_1, \dots, x_m, y; f_1, \dots, f_n)$ such that for all $\bar{a} \in \omega^m$, we have

$$\bar{a} \in X \ \Leftrightarrow\ (\forall f_1)(\exists f_2)(\forall f_3) \cdots (Q f_n)(\overline{Q} y) R(\bar{a}, y; f_1, \dots, f_n), \qquad (2)$$

where the last quantifiers are as follows:

$$Q = \begin{cases} \forall, & \text{if } n \text{ is odd,} \\ \exists, & \text{if } n \text{ is even;} \end{cases} \qquad \overline{Q} = \begin{cases} \exists, & \text{if } Q = \forall, \\ \forall, & \text{if } Q = \exists. \end{cases}$$

Mutatis mutandis, a Σ^1_n set $X \subseteq \omega^m$ can be represented via the following form:

$$\bar{a} \in X \ \Leftrightarrow\ (\exists f_1)(\forall f_2)(\exists f_3) \cdots (Q f_n)(\overline{Q} y) R(\bar{a}, y; f_1, \dots, f_n).$$

Let $\Gamma \in \{\Sigma^1_n, \Pi^1_n : n \geq 1\}$. By $\breve{\Gamma}$ we denote the dual class:

$$\breve{\Gamma} = \begin{cases} \Pi^1_n, & \text{if } \Gamma = \Sigma^1_n, \\ \Sigma^1_n, & \text{if } \Gamma = \Pi^1_n. \end{cases}$$

The next lemma will be useful for transferring various results from a class Γ into its dual $\breve{\Gamma}$.

Definition 3.1. Suppose that $\mathcal{S}$ is a countable family of subsets of ω. By $Dual(\mathcal{S})$ we denote the following family:

$$Dual(\mathcal{S}) := \{A : \ \omega \setminus A \in \mathcal{S}\}.$$

Lemma 3.1. *Suppose that* $\Gamma \in \{\Sigma^1_n, \Pi^1_n\}$. *Then the operator* Dual: $\mathcal{S} \mapsto Dual(\mathcal{S})$ *is a bijection from*

$$\{\mathcal{S} \subset P(\omega) : \mathcal{S} \textit{ is } \Gamma\textit{-computable}\} \textit{ onto } \{\mathcal{T} \subset P(\omega) : \mathcal{T} \textit{ is } \breve{\Gamma}\textit{-computable}\}.$$

Furthermore, the semilattices $\mathcal{R}_\Gamma(\mathcal{S})$ *and* $\mathcal{R}_{\breve{\Gamma}}(Dual(\mathcal{S}))$ *are isomorphic.*

Proof Sketch. Given a Γ-computable numbering ν of a family $\mathcal{S}$, we define a numbering ν^{Dual} of the family $Dual(\mathcal{S})$ as follows: for $k \in \omega$,

$$\nu^{Dual}(k) := \omega \setminus \nu(k).$$

Clearly, the set $G_{\nu^{Dual}}$ is the complement of $G_\nu \in \Gamma$, and hence, the numbering ν^{Dual} is $\breve{\Gamma}$-computable. Furthermore, it is not hard to see that for arbitrary numberings ν and μ, we have:

$$\nu \leq \mu \ \Leftrightarrow \ \nu^{Dual} \leq \mu^{Dual}.$$

This implies that the structures $\mathcal{R}_\Gamma(\mathcal{S})$ and $\mathcal{R}_{\breve{\Gamma}}(Dual(\mathcal{S}))$ are isomorphic.

Note the following: if $\mathcal{S} \neq \mathcal{S}^*$, then $Dual(\mathcal{S}) \neq Dual(\mathcal{S}^*)$. Moreover, $Dual(Dual(\mathcal{S})) = \mathcal{S}$. These facts are enough to finish the proof. □

3.1. *Universal numberings*

Here we give a brief discussion on universal numberings in the analytical hierarchy. We emphasize that the results of this subsection *do not rely on* additional set-theoretic assumptions.

Let n be a non-zero natural number, and let $\Gamma \in \{\Pi^1_n, \Sigma^1_n\}$. We say that a Γ-computable numbering ν of a family $\mathcal{S}$ is *universal* if ν induces the greatest element in the semilattice $\mathcal{R}_\Gamma(\mathcal{S})$.

Proposition 3.1 (Kleene, see XIX in Ref.[30]). *Let* $\mathcal{S}$ *be the family of all* Γ *sets. There is a universal* Γ*-computable numbering of* $\mathcal{S}$.

Proof. Here we give a formal proof of the fact, so that a reader could get familiar with the techniques. We discuss only the case when $\Gamma = \Pi^1_n$. The proof for the Σ^1_n case can be obtained in a similar way, mutatis mutandis.

For a natural number e, set

$$x \in \nu(e) \ \Leftrightarrow \ (\forall f_1)(\exists f_2)(\forall f_3)\cdots(Qf_n)(\overline{Q}s)U(x, s; f_1, f_2, \ldots, f_n),$$

where a recursive predicate U is defined as follows:

- If n is odd, then $U(x, s; f_1, \dots, f_n)$ holds if and only if $\varphi_{e,\langle s\rangle_0}^{f_1 \oplus \cdots \oplus f_n}(x, \langle s\rangle_1) = 1$.
- If n is even, then $U(x, s; f_1, \dots, f_n)$ is true if and only if $\varphi_{e,\langle s\rangle_0}^{f_1 \oplus \cdots \oplus f_n}(x, \langle s\rangle_1)$ is either undefined or equal to one.

Clearly, the set G_ν is Π^1_n and hence, the numbering ν is Π^1_n-computable.

Let $\vec{F}$ denote an oracle $f_1 \oplus \cdots \oplus f_n$, where each f_i belongs to ω^ω. Note the following key property of the relation U: for any oracle $\vec{F}$ and any number x, we have:

- If n is odd, then the condition $\exists s U(x, s; f_1, \dots, f_n)$ is equivalent to $\exists y(\varphi_e^{\vec{F}}(x, y) \downarrow = 1)$.
- If n is even, then $\forall s U(x, s; f_1, \dots, f_n)$ holds iff for every y, either $\varphi_e^{\vec{F}}(x, y) \uparrow$ or $\varphi_e^{\vec{F}}(x, y) \downarrow = 1$.

Let $X \subseteq \omega$ be an arbitrary Π^1_n set. Choose a normal form from Eq. (2) for the set X, and fix a number i_0 witnessing the recursiveness of R from Eq. (2). Then the key property of U implies that $\nu(i_0) = X$. Hence, we deduce that ν is a numbering of the family of *all* Π^1_n sets.

Now let μ be an arbitrary Π^1_n-computable numbering of $\mathcal{S}$. We need to show that $\mu \leq \nu$. Fix an index j_0 witnessing the recursiveness of the set $G_\mu \subseteq \omega$. Recall that this implies the following: $x \in \mu(n)$ iff

$$(\forall f_1)(\exists f_2)(\forall f_3) \cdots (Q f_n)(\overline{Q} y)(\varphi_{j_0}^{\vec{F}}(\langle n, x\rangle, y) = 1).$$

By a relativized *s-m-n* Theorem, there is a computable function $g(u, v)$ such that

$$(\forall \vec{F})(\forall u)(\forall n)(\forall x)(\forall y)[\varphi_u^{\vec{F}}(\langle n, x\rangle, y) = \varphi_{g(u,n)}^{\vec{F}}(x, y)].$$

Thus, the key property of the predicate U implies that $\mu(n) = \nu(g(j_0, n))$ for all n. Hence, μ is reducible to ν. Proposition 3.1 is proved. □

Notice that in the proof above, we never use the fact that the numbering μ gives indices for *all* Π^1_n sets. Therefore, we deduce the following:

Corollary 3.1. *Let* $\Gamma \in \{\Pi^1_n, \Sigma^1_n\}$. *There is a* Γ*-computable numbering* ν_U *such that an arbitrary* Γ*-computable numbering* μ *is reducible to* ν_U.

3.2. *Projective determinacy and its consequences*

Tanaka[41] developed recursion theory for subsets of ω, belonging to the levels of analytical hierarchy, under the assumption of *Projective Determinacy* (**PD**). Here we follow Tanaka's approach: this subsection gives a brief overview of some consequences of **PD**, which will be heavily used in the proofs of our results.

First, we discuss the necessary set-theoretic background, our exposition mainly follows Refs.[25,34].

With each set $A \subseteq \omega^\omega$, one associates a two-person game $G = G(A)$ as follows. Players I and II alternatively choose natural numbers: player I chooses a_0, then II chooses b_0, then I chooses a_1, then II chooses b_1, and so on. The game G ends after ω steps. If the resulting sequence

$$f := (a_0, b_0, a_1, b_1, \ldots)$$

belongs to A, then player I wins. If $f \notin A$, then II wins.

The game G is a game of perfect information: before I choose a_{n+1}, she is allowed to see the tuple $(a_0, b_0, \ldots, a_n, b_n)$; and similarly with II. A *strategy* for the player I is a function σ, which maps tuples of even length to natural numbers. A strategy σ is *winning* if I always wins by following σ. In a similar way, one introduces the notion of a winning strategy for player II. The game $G(A)$ is *determined* if one of the players has a winning strategy.

The *Axiom of Determinacy* (**AD**) states that for every $A \subseteq \omega^\omega$, the game $G(A)$ is determined.

Now we recall the definition of *projective hierarchy* for a Polish space $\mathcal{X}$. Recall that *Baire space* is the set ω^ω, endowed with the natural topology.

A subset $A \subseteq \mathcal{X}$ is *analytic* if either $A = \emptyset$, or there is a continuous map $f\colon \omega^\omega \to \mathcal{X}$ such that $range(f) = A$.

For a non-zero natural number n, the boldface classes $\mathbf{\Sigma}^1_n$ and $\mathbf{\Pi}^1_n$ (for the space $\mathcal{X}$) can be introduced as follows:

- A set $A \subseteq \mathcal{X}$ is $\mathbf{\Sigma}^1_1$ if A is analytic.
- A is $\mathbf{\Pi}^1_1$ if its complement $\mathcal{X} \setminus A$ is analytic.
- A is $\mathbf{\Sigma}^1_{n+1}$ if there is a $\mathbf{\Pi}^1_n$ set $B \subseteq \mathcal{X} \times \omega^\omega$ such that

$$A = \{a : (\exists f \in \omega^\omega)[(a, f) \in B]\}.$$

- A is $\mathbf{\Pi}^1_{n+1}$ if its complement is $\mathbf{\Sigma}^1_{n+1}$.

A set $A \subseteq \mathcal{X}$ is *projective* if A belongs to $\bigcup_{1\leq n<\omega} \mathbf{\Sigma}^1_n = \bigcup_{1\leq n<\omega} \mathbf{\Pi}^1_n$.

The axiom of *Projective Determinacy* (**PD**) states that for every projective set $A \subseteq \omega^\omega$, the game $G(A)$ is determined.

For a detailed discussion of **AD** and **PD**, the reader is referred to, e.g., Refs.[25,34].

We follow Refs.[2,41] and use the following notations: for a natural number k,

- E^1_{2k+1} (pronounced "Epsilon$^1_{2k+1}$") is the (lightface) class Π^1_{2k+1}, and Υ^1_{2k+1} is the class Σ^1_{2k+1};
- $E^1_{2k+2} = \Sigma^1_{2k+2}$ and $\Upsilon^1_{2k+2} = \Pi^1_{2k+2}$.

3.2.1. *The prewellordering property*

A binary relation $\preceq$ on a set S is a *prewellordering* if $\preceq$ is:

- transitive;
- reflexive;
- connected, i.e. $x \preceq y$ or $y \preceq x$ for all $x, y \in S$;
- wellfounded, i.e. S does not contain infinite descending chains $x_0 \succ x_1 \succ x_2 \succ \cdots$.

A *norm* on a set S is an arbitrary function which maps S into the ordinals. Given a norm φ on S, one can associate with φ the following prewellordering:

$$x \preceq^{\varphi} y \;\Leftrightarrow\; \varphi(x) \leq_{Ord} \varphi(y),$$

where $\leq_{Ord}$ is the standard order on ordinals.

Suppose that $S \subseteq \omega$, and $\varphi\colon S \to \lambda$ is a norm on S. The map φ is called a Γ-*norm* if there are binary relations $\preceq^{\varphi}_{\Gamma}$ and $\preceq^{\varphi}_{\breve{\Gamma}}$ on ω such that $\preceq^{\varphi}_{\Gamma}$ belongs to Γ, $\preceq^{\varphi}_{\breve{\Gamma}}$ belongs to $\breve{\Gamma}$, and for every $y \in \omega$,

$$\begin{gathered} \text{if } y \in S, \text{ then for any } x \in \omega, \\ [x \in S \,\&\, \varphi(x) \leq_{Ord} \varphi(y)] \;\Leftrightarrow\; x \preceq^{\varphi}_{\Gamma} y \;\Leftrightarrow\; x \preceq^{\varphi}_{\breve{\Gamma}} y. \end{gathered} \tag{3}$$

A class Γ has the *prewellordering property* (or Γ is *normed*) if every set $S \in \Gamma$ admits a Γ-norm.

The following result is a consequence of the Prewellordering Theorem of Ref.[2]. For a more detailed discussion, see Ref.[33] and Corollary 6B.2 of Ref.[34].

Theorem 3.1 (Addison and Moschovakis[2]). *Assume* ***PD***. *Let n be a non-zero natural number. The class E^1_n has the prewellordering property.*

Remark 3.1. For the classes Π^1_1 and Σ^1_2, this result holds without assuming **PD**. See Theorems XXIII and XXXVIII of Chap. 16 in Ref.[38]; Theorems 4B.2 and 4B.3 of Ref.[34].

Given an arbitrary E^1_n-computable numbering $\nu\colon \omega \to P(\omega)$, we define a relation $\sqsubseteq^{\nu} \subseteq \omega^2 \times \omega^2$ as follows.

Since the set $G_\nu = \{\langle k, x\rangle : x \in \nu(k)\}$ is E^1_n, by Theorem 3.1, one can choose a E^1_n-norm φ mapping G_ν into some ordinal λ. Fix binary relations $\preceq^\varphi_\Gamma$ and $\preceq^\varphi_{\check{\Gamma}}$, where $\Gamma = E^1_n$, witnessing that φ is a E^1_n-norm.

For natural numbers k, m, x, y, we say that $(k, x) \sqsubseteq^\nu (m, y)$ if

$$(k = m) \;\&\; [(\langle k, x\rangle \preceq^\varphi_\Gamma \langle m, y\rangle \;\&\; \langle m, y\rangle \not\preceq^\varphi_\Gamma \langle k, x\rangle) \vee \\ (\langle k, x\rangle \preceq^\varphi_\Gamma \langle m, y\rangle \;\&\; \langle m, y\rangle \preceq^\varphi_\Gamma \langle k, x\rangle \;\&\; x \leq_\omega y)].$$

Informally speaking, the result below introduces the "building blocks" of our constructions: The sets $\widehat{[x]}_{\nu(k)}$ allow us to transfer some results (already known for, say, Σ^0_2-computable numberings) into our setting.

Lemma 3.2 (Main Property of $\sqsubseteq^\nu$). *Assume* ***PD****. Suppose that ν is a E^1_n-computable numbering, and $k \in \omega$. Then:*

(i) The relation $\preceq^{\nu(k)} := \{(x, y) : x, y \in \nu(k),\ (k, x) \sqsubseteq^\nu (k, y)\}$ is a wellordering on the set $\nu(k)$.

(ii) For a number $x \in \nu(k)$, the set

$$\widehat{[x]}_{\nu(k)} := \{z \in \omega : (k, z) \sqsubseteq^\nu (k, x)\}$$

is a Δ^1_n subset of $\nu(k)$. Furthermore, the formulas witnessing the Δ^1_n-ness do not depend on the choice of k and x.

Proof. (i) The definition of the relation $\sqsubseteq^\nu$ implies that for arbitrary numbers x, y from $\nu(k)$, the condition $x \preceq^{\nu(k)} y$ holds iff either $\varphi(\langle k, x\rangle) \lneq_{Ord} \varphi(\langle k, y\rangle)$, or $\varphi(\langle k, x\rangle) = \varphi(\langle k, y\rangle)$ & $x \leq_\omega y$. Hence, the map

$$\psi\colon x \mapsto (x, \varphi(\langle k, x\rangle))$$

induces an isomorphic embedding from $\nu(k)$ into the ordinal $\omega\cdot\lambda$. Therefore, we deduce that the poset $(\nu(k); \preceq^{\nu(k)})$ is well-ordered.

(ii) Since $\langle k, x\rangle \in G_\nu$, for an arbitrary $z \in \widehat{[x]}_{\nu(k)}$, we have $\langle k, z\rangle \preceq^\varphi_\Gamma \langle k, x\rangle$ and hence, by the definition of a Γ-norm, $\langle k, z\rangle \in G_\nu$. In other words, we showed that $\widehat{[x]}_{\nu(k)} \subseteq \nu(k)$.

Eq. (3) implies that the following conditions are equivalent:

(a) $z \preceq^{\nu(k)} x$;

(b) $(\langle k, z\rangle \preceq^\varphi_\Gamma \langle k, x\rangle$ and $\langle k, x\rangle \not\preceq^\varphi_{\check{\Gamma}} \langle k, z\rangle)$, or $(\langle k, z\rangle \preceq^\varphi_\Gamma \langle k, x\rangle$ and $\langle k, x\rangle \preceq^\varphi_\Gamma \langle k, z\rangle$ and $z \leq_\omega x)$;

(c) $(\langle k, z\rangle \preceq^\varphi_{\check{\Gamma}} \langle k, x\rangle$ and $\langle k, x\rangle \not\preceq^\varphi_\Gamma \langle k, z\rangle)$, or $(\langle k, z\rangle \preceq^\varphi_{\check{\Gamma}} \langle k, x\rangle$ and $\langle k, x\rangle \preceq^\varphi_{\check{\Gamma}} \langle k, z\rangle$ and $z \leq_\omega x)$.

Condition (b) is logically equivalent to a E_n^1 formula, and Condition (c) is equivalent to a Υ_n^1 formula. Clearly, the formulas do not depend on the choice of k and x — they only depend on the choice of the relations $\preceq_\Gamma^\varphi$ and $\preceq_\Gamma^\varphi$. Lemma 3.2 is proved. □

3.2.2. *Reduction principle*

Suppose that $\Gamma \in \{\Sigma_n^1, \Pi_n^1 : n \geq 1\}$. The class Γ satisfies the *reduction principle* if for every pair $A, B \in \Gamma$, there are $A^*, B^* \in \Gamma$ such that:

$$A^* \subseteq A, \quad B^* \subseteq B,$$
$$A^* \cup B^* = A \cup B, \quad A^* \cap B^* = \emptyset.$$

Theorem 3.2 (Addison and Moschovakis[2]; Martin[32]). *Assume* ***PD****. Let n be a non-zero natural number. The class E_n^1 satisfies the reduction principle.*

Remark 3.2. The classes Π_1^1 and Σ_2^1 satisfy reduction principle, without assuming **PD** (Addison[1]).

Note that the reduction principle for E_n^1 (as formulated above) follows from Theorem 3.1, see Exercise 4B.10 and Corollary 6B.2 of Ref.[34].

4. Rogers semilattices for finite families

For the sake of convenience, from now on, we will use the following notation:

$$\mathcal{R}_n^1(\mathcal{S}) := \mathcal{R}_{E_n^1}(\mathcal{S}).$$

The section discusses elementary properties of the semilattices $\mathcal{R}_n^1(\mathcal{S})$, where $\mathcal{S}$ is finite. As a warm-up, we apply standard numbering-theoretic techniques (see, e.g., Ref.[6]) to prove that $\mathcal{R}_n^1(\mathcal{S})$ is a distributive upper semilattice (Proposition 4.1). We also show that if $\mathcal{S}$ contains more than one element, then $\mathcal{R}_n^1(\mathcal{S})$ is infinite and cannot be a lattice (Proposition 4.2). Note that these proofs *do not require* additional set-theoretic assumptions. After that, we obtain a criterion for when $\mathcal{R}_n^1(\mathcal{S})$ has the greatest element (Theorem 4.1). The proof of this criterion already heavily employs the consequences of **PD** discussed in Sec. 3.2. After that, the developed techniques are used to prove the following: If $\mathcal{R}_n^1(\mathcal{S})$ has no greatest element, then it is upwards dense (Theorem 4.2).

Before proceeding to main results, we observe the following very simple fact:

Remark 4.1. Let $\mathcal{S}$ be a non-empty finite family of E^1_n sets. Then $\mathcal{S}$ is E^1_n-computable, and the semilattice $\mathcal{R}^1_n(\mathcal{S})$ contains the least element.

Proof. Suppose that $\mathcal{S} = \{B_0, B_1, \ldots, B_m\}$. Then the numbering

$$\mu(k) := \begin{cases} B_k, & \text{if } k \leq m, \\ B_0, & \text{otherwise}, \end{cases}$$

is a E^1_n-computable numbering of $\mathcal{S}$.

Assume that ν is an arbitrary E^1_n-computable numbering of $\mathcal{S}$. Choose indices b_i, $i \leq m$, with $\nu(b_i) = B_i$. Clearly, a computable function

$$f(k) := \begin{cases} b_k, & \text{if } k \leq m, \\ b_0, & \text{otherwise}, \end{cases}$$

provides a reduction from μ into ν. □

Recall that an upper semilattice $\mathcal{A} = (A; \leq, \vee)$ is *distributive* if for all $b, a_0, a_1 \in \mathcal{A}$, the following holds:

$$(b \leq a_0 \vee a_1) \Rightarrow \exists b_0 \exists b_1 [(b = b_0 \vee b_1) \,\&\, (b_0 \leq a_0) \,\&\, (b_1 \leq a_1)].$$

The next two propositions establish general facts about Rogers E^1_n-semilattices of finite families, note that these facts do not depend on **PD**.

Proposition 4.1. *Suppose that $\mathcal{S} = \{A_0, A_1, \ldots, A_m\}$ is a finite family of E^1_n sets. Then the upper semilattice $\mathcal{R}^1_n(\mathcal{S})$ is distributive.*

Proof. Let μ, ν_0, and ν_1 be E^1_n-computable numberings of $\mathcal{S}$. Suppose that a computable function $f(x)$ reduces μ to $\nu_0 \oplus \nu_1$. Consider computable sets

$$R_0 = \{x : f(x) \text{ is even}\}, \quad R_1 = \{x : f(x) \text{ is odd}\}.$$

First, assume that one of these sets, say, R_1 is empty. Then μ is reducible to ν_0, and one can define $\mu_0 := \mu$, and

$$\mu_1(k) := \begin{cases} A_k, & \text{if } k \leq m, \\ A_0, & \text{if } k > m. \end{cases}$$

It is not difficult to see that $\mu \equiv \mu_0 \oplus \mu_1$, $\mu_0 \leq \nu_0$, and $\mu_1 \leq \nu_1$.

Therefore, from now on, we may assume that both R_0 and R_1 are non-empty. For $i \in \{0, 1\}$, fix a total computable function g_i with $range(g_i) = R_i$. Define a E^1_n-computable numbering $\xi_i := \mu \circ g_i$. It is not hard to show (see, e.g., Proposition 3.1 of Ref.[6]) that $\xi_i \leq \nu_i$ and $\mu \equiv \xi_0 \oplus \xi_1$. Note that in general, ξ_i indexes *not all* elements of $\mathcal{S}$.

For $i \in \{0,1\}$, we define

$$\mu_i(k) := \begin{cases} A_k, & \text{if } k \leq m, \\ \xi_i(k-m-1), & \text{if } k > m. \end{cases}$$

Clearly, each μ_i is a E^1_n-computable numbering of the family $\mathcal{S}$. On the other hand, it is straightforward to show that $\mu_i \leq \nu_i$ and $\mu \equiv \mu_0 \oplus \mu_1$. Therefore, the semilattice $\mathcal{R}^1_n(\mathcal{S})$ is distributive. This concludes the proof of Proposition 4.1. □

For an E^1_n set A, consider a one-element family $\mathcal{T} := \{A\}$. It is obvious that the family $\mathcal{T}$ has *only one* numbering, and hence, the semilattice $\mathcal{R}^1_n(\mathcal{T})$ is one-element.

The next proposition shows that if a finite family $\mathcal{S}$ contains more than one element, then for this $\mathcal{S}$, one can recast the results of Khutoretskii[28] and Selivanov[40], mentioned in the introduction.

Proposition 4.2. *Let $\mathcal{S} = \{A_0, A_1, \dots, A_m\}$ be a finite family of E^1_n sets, which contains at least two elements. Then the Rogers semilattice $\mathcal{R}^1_n(\mathcal{S})$ is infinite, and it is not a lattice.*

Proof. We follow the proof of Theorem 2.1 in Ref.[24]. Note that $m \geq 1$. Let W be a computably enumerable set such that $W \neq \emptyset$ and $W \neq \omega$. We define a numbering ν^W as follows: for $k \in \omega$, set

$$\nu^W(k) := \begin{cases} A_k, & \text{if } k \leq m-2, \\ A_{m-1}, & \text{if } k \geq m-1 \text{ and } (k-m+1) \in W, \\ A_m, & \text{if } k \geq m-1 \text{ and } (k-m+1) \notin W. \end{cases}$$

It is clear that ν^W is a E^1_n-computable numbering of the family $\mathcal{S}$.

Let μ be a numbering of $\mathcal{S}$ such that $\mu \leq \nu^W$. Fix a computable function f, which reduces μ to ν^W. Then the set $V := \{k : f(k) \in W\}$ is c.e., and it is straightforward to show that $V \leq_m W$ and $\mu \equiv \nu^V$. Consider a principal ideal $\mathcal{I}$, induced inside $\mathcal{R}^1_n(\mathcal{S})$ by the (degree of the) numbering $\nu^{\emptyset'}$. One can show that $\mathcal{I}$ is isomorphic to the upper semilattice of c.e. m-degrees $\mathbf{R}_m$. Since $\mathbf{R}_m$ is not a lattice, the structure $\mathcal{R}^1_n(\mathcal{S})$ also cannot be a lattice. Proposition 4.2 is proved. □

4.1. *Existence of a greatest element*

Now we proceed to investigating the following question (while assuming **PD**): When does the semilattice $\mathcal{R}^1_n(\mathcal{S})$ has the greatest element?

Theorem 4.1 (PD). *Let n be a non-zero natural number. Let $\mathcal{S} = \{A_1, A_2, \ldots, A_m\}$ be a finite family of E^1_n sets. Then the Rogers semilattice $\mathcal{R}^1_n(\mathcal{S})$ has the greatest element if and only if the family $\mathcal{S}$ contains a least element under set-theoretic inclusion.*

Proof. The basic idea is essentially the same as that of Theorem 3.2 in Ref.[5], but we also have to carefully interweave set-theoretic results from Sec. 3.2.

Choose a family of finite sets $\{F_1, F_2, \ldots, F_m\}$ with the following property: for all i and j, we have

$$F_i \subseteq F_j \Leftrightarrow A_i \subseteq A_j \Leftrightarrow F_i \subseteq A_j.$$

We fix a universal E^1_n-computable numbering ν_U from Corollary 3.1. For the sake of brevity, the relation $\sqsubseteq^{\nu^U}$ from Lemma 3.2 will be denoted by $\sqsubseteq$.

($\Leftarrow$). Suppose that A_1 is the least element inside $\mathcal{S}$. Without loss of generality, we may assume that $F_1 = \emptyset$.

Before giving a formal construction, we outline the intuition behind our proof. For starters, we sketch a *classical* argument: Assume that a family $\mathcal{T}$ contains precisely the following sets:

$$\emptyset, \{0\}, \{1\}, \{0,1\}, \{1,2\},$$

and we want to build a universal Σ^0_1-computable numbering of $\mathcal{T}$.

In order to achieve this, first, we choose a universal Σ^0_1-computable numbering ν of the family of *all* c.e. sets. Note that, similarly to Corollary 3.1, *any* Σ^0_1-computable numbering (of any family) is reducible to ν.

We describe an effective procedure that transforms the numbering ν into a Σ^0_1-computable numbering ξ, which indexes *only* elements from $\mathcal{T}$. Fix an effective uniform approximation

$$\nu(k) = \bigcup_{s \in \omega} \nu^s(k),$$

where every $\nu^s(k)$ is a finite set, and $\nu^s(k) \subseteq \nu^{s+1}(k)$. As usual, we may assume that the difference $\nu^{s+1}(k) \setminus \nu^s(k)$ contains at most one element.

The desired procedure works in a pretty straightforward, "dynamic" fashion:

(a) While neither 0, nor 1 belongs to $\nu^s(k)$, we just put $\xi^s(k) := \emptyset$. Assume that s_0 is the least step such that (precisely) one of the elements 0 or 1 appears in $\nu^{s_0}(k)$.

(b.1) If $0 \in \nu^{s_0}(k)$, then define $\xi^{s_0}(k) := \{0\}$. Wait for a first step $s_1 > s_0$ with $1 \in \nu^{s_1}(k)$. While waiting, do not change the (approximation of the) set $\xi(k)$. If such a step appears, then set $\xi(k) = \xi^{s_1}(k) := \{0, 1\}$.

(b.2) Otherwise, we have $1 \in \nu^{s_0}(k)$. Put $\xi^{s_0}(k) := \{1\}$. Wait for a first step $s_2 > s_0$ such that one of the elements 0 or 2 belongs to $\nu^{s_2}(k)$. While waiting, don't change $\xi(k)$. When such a step s_2 is found, denote the element from $\{0, 2\} \cap \nu^{s_2}(k)$ as v, and define $\xi(k) = \xi^{s_2}(k) := \{0, v\}$.

Clearly, the described procedure is uniform in k. Moreover, it builds a Σ^0_1-computable numbering ξ, and for each k, the constructed $\xi(k)$ is an element from $\mathcal{T}$. The key property of the construction is provided by the following simple observation: if $\nu(k)$ is already an element of $\mathcal{T}$, then $\xi(k)$ is *equal* to $\nu(k)$.

Now let μ be an arbitrary Σ^0_1-computable numbering of $\mathcal{T}$. Since the numbering ν is universal, one can choose a computable function $f(x)$ such that $\mu(k) = \nu(f(k))$ for all numbers k. The key property of the construction implies that $\mu(k) = \nu(f(k)) = \xi(f(k))$. In other words, $\mu \leq \xi$, and ξ induces the greatest element of the Rogers semilattice for the Σ^0_1-computable family $\mathcal{T}$.

This concludes the description of the classical argument. The argument *cannot* be readily transferred to the E^1_n setting: roughly speaking, we do not know what is an effective approximation of a given E^1_n-computable numbering ν. Nevertheless, this obstacle can be circumvented as follows — a careful analysis of the construction above reveals that the *only* important question we need to address is the following:

What does one mean, when she says: "The number 0 appears *earlier* than the number 1, in an approximation of $\nu(k)$"?

And we *can* give a precise answer to this question: Zero appears earlier than one iff there is a number $x \in \nu(k)$ such that $0 \in \widehat{[x]}_{\nu(k)}$ and $1 \notin \widehat{[x]}_{\nu(k)}$. In other words, the wellorder $(\nu(k); \preceq^{\nu(k)})$ provided by Lemma 3.2 allows us to "replace" the stages s_0, s_1, s_2, used in the classical construction, by elements x_0, x_1, x_2 belonging to the set $\nu(k)$ itself. The formal details of this (quite informal) idea are elaborated below.

Consider a set $I := \{2, 3, \dots, m\}$, note that here we do not include 1 into I. We introduce a partial order $\trianglelefteq$ on I as follows. For $i, j \in I$, we say that $i \trianglelefteq j$ iff $A_i \subseteq A_j$.

For a number $i \in I$, its upper cone is the following set:

$$\widetilde{(i)} := \{j \in I : j \trianglerighteq i,\ j \neq i\}.$$

Notice that i itself does not belong to the cone $\widetilde{(i)}$.

Let $C = \{i_0 \lhd i_1 \lhd \cdots \lhd i_t\}$ be a maximal (under set-inclusion) increasing chain inside the poset $(I; \unlhd)$. We define the following formulas:

$$\psi_0^C(k) := \exists x \Big[x \in \nu^U(k) \,\&\, \bigwedge_{y \in F_{i_0}} (k,y) \sqsubseteq (k,x) \,\&$$

$$\bigwedge_{j \text{ is minimal inside } I;\, j \neq i_0} \ \bigvee_{z \in F_j} (k,z) \not\sqsubseteq (k,x) \Big];$$

$$\psi_{l+1}^C(k) := \psi_l^C(k) \,\&\, \exists x \Big[x \in \nu^U(k) \,\&\, \bigwedge_{y \in F_{i_{l+1}}} (k,y) \sqsubseteq (k,x) \,\&$$

$$\bigwedge_{j \text{ is minimal inside } \widetilde{(i_l)};\, j \neq i_{l+1}} \ \bigvee_{z \in F_j} (k,z) \not\sqsubseteq (k,x) \Big];$$

$$\psi^C(x,k) := \bigvee_{l \leq t} [\psi_l^C(k) \,\&\, x \in A_{i_l}].$$

The main property of the relation $\sqsubseteq$ (Lemma 3.2) implies that each of these formulas is logically equivalent to a E_n^1 condition.

We define a new numbering ξ as follows: $x \in \xi(k)$ iff

$$(x \in A_1) \text{ or } \bigvee_{C \text{ is a maximal chain in } I} \psi^C(x,k).$$

Clearly, the set G_ξ belongs to E_n^1 and thus, the numbering ξ is E_n^1-computable. We establish the following important property of ξ:

Claim 4.1. *For any k, the set $\xi(k)$ belongs to the family $\mathcal{S}$. Moreover, if $\nu^U(k) \in \mathcal{S}$, then $\xi(k) = \nu^U(k)$.*

Proof. Let k be an arbitrary natural number. Without loss of generality, we may assume that $\xi(k) \neq A_1$. It is sufficient to prove the following fact: There is a maximal chain $C = \{i_0 \lhd i_1 \lhd \cdots \lhd i_t\}$ and a number $r \leq t$ such that $\xi(k) = \bigcup_{l \leq r} A_{i_l} = A_{i_r}$. We build the desired chain C step-by-step.

Since $\xi(k) \neq A_1$, there is a chain D_0 inside I such that $\psi_0^{D_0}(k)$ holds. We define i_0 as the least element of D_0. Recall that Lemma 3.2 shows the following: if $x \in \nu^U(k)$ and $(k,y) \sqsubseteq (k,x)$, then y also belongs to $\nu^U(k)$. Thus, we deduce that $F_{i_0} \subseteq \nu^U(k)$ and $A_{i_0} \subseteq \xi(k)$.

The relation $\preceq^{\nu(k)}$ is a wellordering on $\nu(k)$. This implies that there is the $\preceq^{\nu(k)}$-least number $x_0 \in \nu(k)$ such that $F_j \subseteq \widehat{[x_0]}_{\nu(k)}$ for some $\unlhd$-minimal $j \in I$. Since $\psi_0^{D_0}(k)$ is true, this particular $\unlhd$-minimal j must be equal to i_0. From this, we deduce the following: if a maximal chain D' starts with an element j' not equal to i_0, then $\psi^{D'}(x,k)$ is false for all x. Such chains D' can be omitted from further consideration.

Now assume that i_l has been already defined, and we know the following:

(a) $i_0 \lhd i_1 \lhd \cdots \lhd i_l$;
(b) $F_{i_l} \subseteq \nu^U(k)$ and $A_{i_l} \subseteq \xi(k)$;
(c) For each $u \leq l$, if a maximal chain D' does not start with $i_0 \lhd i_1 \lhd \cdots \lhd i_u$, then $\psi_q^{D'}(k)$ is false for all $q \geq u$.

Consider the following cases:

Case 1. Assume that $C := \{i_0 \lhd i_1 \lhd \cdots \lhd i_l\}$ is already a maximal chain inside I. Then the items (b) and (c) together imply that $\xi(k) = \bigcup_{u \leq l} A_{i_u} = A_{i_l}$, and this finishes the construction.

Case 2. Suppose that $i_0 \lhd \cdots \lhd i_l$ is not a maximal chain.

Case 2.1. Assume that for every $j \rhd i_l$, we have $F_j \not\subseteq \nu^U(k)$. Then Lemma 3.2 implies that for any maximal chain D starting with $i_0 \lhd \cdots \lhd i_l$, the formula $\psi_{l+1}^D(k)$ is false. Hence, one obtains that $\xi(k) = A_{i_l}$, and the desired C can be chosen as an arbitrary maximal chain beginning with $i_0 \lhd \cdots \lhd i_l$.

Case 2.2. Suppose that there is $j \rhd i_l$ with $F_j \subseteq \nu^U(k)$. Since $\preceq^{\nu(k)}$ is a wellordering, there is the $\preceq^{\nu(k)}$-least number $x_{l+1} \in \nu(k)$ such that $F_{j^*} \subseteq \widehat{[x_{l+1}]_{\nu(k)}}$ for some $\unlhd$-minimal j^* from $\widetilde{(i_l)}$. Furthermore, this number j^* is uniquely determined, and for any maximal chain D starting with $i_0 \lhd \cdots \lhd i_l \lhd j^*$, the formula $\psi_{l+1}^D(k)$ is true. We put $i_{l+1} := j^*$ and proceed to finding i_{l+2}. Clearly, the following holds:

- $i_{l+1} \rhd i_l$;
- $F_{i_{l+1}} \subseteq \nu^U(k)$ and $A_{i_{l+1}} \subseteq \xi(k)$;
- For every maximal chain D' not beginning with $i_0 \lhd \cdots \lhd i_l \lhd i_{l+1}$, the formulas $\psi_q^{D'}$, $q \geq l+1$, are false.

Since the poset I is finite, the described construction finishes after finitely many iterations, and produces the desired maximal chain C and number i_r such that $\xi(k) = A_{i_r}$. In particular, every $\xi(k)$ belongs to $\mathcal{S}$.

Now assume that $\nu^U(k) = A_j$ for some $j \leq m$. Clearly, if $j = 1$, then $\xi(k)$ is also equal to A_1. Hence, we assume that $j \geq 2$. In this particular case, the construction above finishes only when one of the following two conditions is satisfied:

(1) We found a $\unlhd$-maximal i_l with $F_{i_l} \subseteq \nu^U(k)$. Then, clearly, $j = i_l$ and

$\xi(k) = \nu^U(k)$.

(2) We found a number i_l such that $F_{i_l} \subseteq \nu^U(k)$, but for every $q \rhd i_l$, the set F_q is not a subset of $\nu^U(k)$. We deduce that $A_{i_l} \subseteq A_j$, and for any q, the condition $A_q \supsetneq A_{i_l}$ implies $A_q \neq A_j$. Hence, $j = i_l$ and $\xi(k) = A_{i_l} = \nu^U(k)$.

Claim 4.1 is proved. □

Recall that the numbering ν^U is universal. Thus, for any E^1_n-computable numbering μ of the family $\mathcal{S}$, there is a computable function $f(x)$ such that $\mu(k) = \nu^U(f(k))$ for all k. Claim 4.1 implies that $\mu(k) = \nu^U(f(k)) = \xi(f(k))$, and therefore, ξ is a universal E^1_n-computable numbering of the family $\mathcal{S}$.

<u>$(\Rightarrow)$.</u> Suppose that the family $\mathcal{S}$ has no least element under $\subseteq$. Without loss of generality, we may assume that $A_1, A_2, \dots, A_l$ (where $2 \leq l \leq m$) are all $\subseteq$-minimal elements of $\mathcal{S}$.

In order to prove the desired fact, it is sufficient to obtain the following: Given an arbitrary E^1_n-computable numbering ν of the family $\mathcal{S}$, one can construct a E^1_n-computable numbering μ of $\mathcal{S}$ such that $\mu \not\leq \nu$.

Fix indices p_i, $1 \leq i \leq l$, such that $\nu(p_i) = A_i$. Let

$$p_{\sigma(i)} := \begin{cases} p_{i+1}, & \text{if } i < l, \\ p_1, & \text{if } i = l. \end{cases}$$

For each non-zero $i \leq l$, consider a E^1_n set

$$Q_i := \{k \in \omega : F_i \subseteq \nu(k)\}.$$

Clearly, $\bigcup_{i \leq l} Q_i$ is equal to ω (recall that we picked *all* $\subseteq$-minimal elements of $\mathcal{S}$). By the reduction principle (Theorem 3.2), there are pairwise disjoint E^1_n sets R_i, $i \leq l$, such that

$$R_i \subseteq Q_i \text{ and } \bigcup_{i \leq l} R_i = \omega.$$

Notice that in particular, every R_i is a Δ^1_n set.

We define a total function $f \colon \omega \to \{p_i : 1 \leq i \leq l\}$. Set

$$f(k) := \begin{cases} p_{\sigma(i)}, & \text{if } \varphi_k(k)\downarrow \text{ and } \varphi_k(k) \in R_i, \\ p_1, & \text{if } \varphi_k(k)\uparrow. \end{cases}$$

Since every $z \in \omega$ belongs to precisely one set R_i, the function f is well-defined. Moreover, the graph of f is a Δ^1_n set (recall that each R_i is Δ^1_n).

We define the desired numbering μ as follows:

$$\xi(k) := \nu(f(k)), \quad \mu := \nu \oplus \xi.$$

First, note that $x \in \xi(k)$ iff $\exists t[f(k) = t \,\&\, x \in \nu(t)]$. This shows that both numberings ξ and μ are E^1_n-computable. Moreover, it is evident that μ indexes precisely the family $\mathcal{S}$.

Towards a contradiction, assume that $\mu \leq \nu$. Then one can choose a total computable function $\varphi_e(x)$ with $\xi = \nu \circ \varphi_e$. Find the number $i \leq l$ such that $\varphi_e(e)$ belongs to R_i. Since $\varphi_e(e) \in R_i \subseteq Q_i$, the set $\xi(e) = \nu(\varphi_e(e))$ must contain F_i. On the other hand, we have $f(e) = p_{\sigma(i)}$ and $\xi(e) = \nu(f(e)) = A_{\sigma(i)}$. The set $A_{\sigma(i)}$ is $\subseteq$-minimal inside $\mathcal{S}$, and hence, $A_{\sigma(i)} \not\supseteq F_i$. This gives a contradiction, therefore, μ is not reducible to ν. Theorem 4.1 is proved. $\square$

A careful analysis of the proof above (see also Sec. 3.2) reveals the following:

Corollary 4.1. *For the classes Π^1_1 and Σ^1_2, Theorem 4.1 holds even without assuming **PD**.*

Furthermore, by applying the operator Dual from Lemma 3.1, one can prove:

Corollary 4.2 (PD). *Let n be a non-zero natural number, and let $\mathcal{S}$ be a finite family of Υ^1_n sets. Then the semilattice $\mathcal{R}_{\Upsilon^1_n}(\mathcal{S})$ has the greatest element if and only if the family $\mathcal{S}$ contains a greatest element under $\subseteq$.*

4.2. *Minimal covers*

The technique developed in the proof of Theorem 4.1 allows us to obtain a further result, which deals with minimal covers in Rogers semilattices.

Let $\mathcal{A} = (A; \leq, \vee)$ be an upper semilattice, and a be an element from $\mathcal{A}$. An element $b \in \mathcal{A}$ is called a *minimal cover* of a if $a < b$ and there is no c with $a < c < b$.

Theorem 4.2 (PD). *Let n be a non-zero natural number. Let $\mathcal{S} = \{A_1, A_2, \ldots, A_m\}$ be a finite family of E^1_n sets such that $\mathcal{S}$ has no least element under $\subseteq$. Then every element from the Rogers semilattice $\mathcal{R}^1_n(\mathcal{S})$ has a minimal cover. In particular, $\mathcal{R}^1_n(\mathcal{S})$ is upwards dense.*

Before giving the proof of the theorem, we obtain the following auxiliary fact:

Proposition 4.3. *Let $\mathcal{T}$ be an arbitrary E^1_n-computable family, and let ν be a E^1_n-computable numbering of $\mathcal{T}$. Suppose that there exists a total function $f\colon \omega \to \omega$ such that:*

- *the graph of f is Δ^1_n, and*
- *$\nu(k) \neq \nu(f(k))$ for all $k \in \omega$.*

Then the (degree of the) numbering ν has a minimal cover inside $\mathcal{R}^1_n(\mathcal{T})$.

Proof. The proof mimics the idea from Theorem 2 of Ref.[10]. Let M be a maximal computably enumerable set. Fix a total computable, injective function g such that $range(g) = M$. Assume that $\overline{M} := \omega \setminus M = \{m_0 <_\omega m_1 <_\omega m_2 <_\omega \cdots\}$.

Define a numbering μ as follows:

$$\mu(k) := \begin{cases} \nu(g^{-1}(k)), & \text{if } k \in M, \\ \nu(0), & \text{if } k = m_e \text{ and } k \notin dom(\varphi_e), \\ \nu(f(\varphi_e(k))), & \text{if } k = m_e \text{ and } k \in dom(\varphi_e). \end{cases}$$

Via standard counting of quantifiers, one can see that the numbering μ is E^1_n-computable: e.g., the condition $x \in \nu(f(\varphi_e(k)))$ is equivalent to a E^1_n formula

$$\exists u \exists v [\varphi_e(k){\downarrow} = u \,\&\, f(u) = v \,\&\, x \in \nu(v)].$$

Clearly, ν is equal to $\mu \circ g$; hence, $\nu \leq \mu$ and μ indexes precisely the family $\mathcal{T}$. Assume that a total computable function φ_e reduces μ to ν. Then $\mu(m_e) = \nu(\varphi_e(m_e))$ and on the other hand, $\mu(m_e) = \nu(f(\varphi_e(m_e))) \neq \nu(\varphi_e(m_e))$, which gives a contradiction. Thus, we deduce that μ is not reducible to ν.

Now assume that ξ is a numbering of $\mathcal{T}$ such that $\nu \leq \xi \leq \mu$. Fix total computable functions h and p such that $\xi = \mu \circ h$ and $\nu = \xi \circ p$. Since the set M is maximal, one of the following two cases holds:

Case 1. The set $range(h) \setminus M$ is finite. Assume that $range(h) \setminus M = \{b_0, b_1, \ldots, b_t\}$ and choose ν-indices a_i, $i \leq t$, such that $\nu(a_i) = \mu(b_i)$. Define a computable function

$$q(k) := \begin{cases} a_i, & \text{if } h(k) = b_i \text{ for some } i \leq t, \\ g^{-1}(h(k)), & \text{otherwise.} \end{cases}$$

The function q is well-defined, and it reduces ξ to ν; hence, $\xi \equiv \nu$.

Case 2. The set $\overline{M} \setminus range(h)$ is finite. Assume that $\overline{M} \setminus range(h) = \{c_0, c_1, \ldots, c_t\}$, and choose ξ-indices d_i, $i \leq t$ such that $\xi(d_i) = \mu(c_i)$. Define a computable function q according to the following rules:

(a) If $k = c_i$, then set $q(k) := d_i$.
(b) Otherwise, find the least step s such that one of the following holds:
   (b.1) There is a number l such that $h(l)[s]\downarrow = k$. Then set $q(k) := l$.
   (b.2) The number k belongs to $M[s]$. Then set $q(k) := p(g^{-1}(k))$.

It is not hard to show that the function q is well-defined, and $\mu = \xi \circ q$, hence $\xi \equiv \mu$.

Summarizing, we showed that there are no numberings strictly between ν and μ, and hence, μ is a minimal cover for ν. Proposition 4.3 is proved. □

Proof of Theorem 4.2. We follow the notations employed by the proof of the direction ($\Rightarrow$) in Theorem 4.1: The sets $A_1, A_2, \dots, A_l$ are chosen as all $\subseteq$-minimal elements from $\mathcal{S}$. Given a E^1_n-computable numbering ν of $\mathcal{S}$, we introduce precisely the same objects p_i, $p_{\sigma(i)}$, Q_i, and R_i as in Theorem 4.1.

We define a total function $f\colon \omega \to \omega$ as follows:

$$f(k) := p_{\sigma(i)}, \text{ if } k \in R_i.$$

Clearly, f is well-defined, and the graph of f is Δ^1_n.

Suppose that $k \in R_i$. Then $A_i \subseteq \nu(k)$. On the other hand, $\nu(f(k)) = \nu(p_{\sigma(i)}) = A_{\sigma(i)}$ and hence, $\nu(f(k)) \not\supseteq A_i$. Thus, $\nu(f(k)) \neq \nu(k)$. By Proposition 4.3, we deduce that the numbering ν has a minimal cover inside $\mathcal{R}^1_n(\mathcal{S})$. Theorem 4.2 is proved. □

5. Rogers semilattices of infinite families

In this section, we discuss elementary properties of the semilattices $\mathcal{R}^1_n(\mathcal{S})$, where $\mathcal{S}$ is infinite. The first property, which differs from the results of Sec. 4, is the following fact: $\mathcal{R}^1_n(\mathcal{S})$ never contains the least element, as witnessed by the result of Dorzhieva:

Proposition 5.1 (Dorzhieva, Corollary 1 in Ref.[14]). *Let n be a non-zero natural number, and let $\mathcal{S}$ be an infinite Π^1_n-computable family. Then the Rogers semilattice $\mathcal{R}_{\Pi^1_n}(\mathcal{S})$ contains infinitely many minimal elements.*

Note that Lemma 3.1 implies the following: if one replaces Π^1_n by Σ^1_n, then Proposition 5.1 still stays true. Moreover, by combining Propositions 4.2 and 5.1, we obtain:

Corollary 5.1. *Let $\mathcal{S}$ be an arbitrary E^1_n-computable family, which contains at least two elements. Then the Rogers semilattice $\mathcal{R}^1_n(\mathcal{S})$ is infinite, and it is not a lattice.*

Furthermore, Dorzhieva extended a result of Podzorov (Theorem 1 of Ref.[35]) and proved the following:

Proposition 5.2 (Dorzhieva, a part of Lemma 1 in Ref.[14]). *Let $\mathcal{S}$ be an infinite Π^1_n-computable family. Then there is a principal ideal $\mathcal{I}$ inside $\mathcal{R}_{\Pi^1_n}(\mathcal{S})$, which is isomorphic to the upper semilattice $\mathcal{E}^* \setminus \{\bot\}$ (i.e. the semilattice of all c.e. sets under set-theoretic almost-inclusion $\subseteq^*$, with the least element omitted).*

We make use of Proposition 5.2 and exhibit a further elementary property, which differs from Sec. 4.

Definition 5.1 (Definition 3.2 and Proposition 3.2 of Ref.[6]). Let $\mathcal{A} = (A; \leq, \vee)$ be an upper semilattice. We say that $\mathcal{A}$ is *weakly distributive* if it satisfies the following: if one adds an external least element $\bot$ to $\mathcal{A}$, then the resulting structure is a distributive upper semilattice. Equivalently, $\mathcal{A}$ is weakly distributive if and only if for all $b, a_0, a_1 \in \mathcal{A}$, the following holds:

$$(b \leq a_0 \vee a_1) \,\&\, (b \nleq a_0) \,\&\, (b \nleq a_1) \;\Rightarrow\; \exists b_0 \exists b_1 [(b = b_0 \vee b_1) \,\&\, (b_0 \leq a_0) \,\&\, (b_1 \leq a_1)].$$

Theorem 5.1. *Let n be a non-zero natural number, and let $\mathcal{S}$ be an infinite E^1_n-computable family. Then the semilattice $\mathcal{R}^1_n(\mathcal{S})$ is not weakly distributive.*

Proof. The proof follows the ideas of Theorem 3.2 in Ref.[6]. We will build three E^1_n-computable numberings μ, ν_0, and ν_1 of the family $\mathcal{S}$, which witness the failure of weak distributivity.

First, we define auxiliary numberings α, β, and γ. Let α be a E^1_n-computable numbering of $\mathcal{S}$ such that the principal ideal $\mathcal{I}$, induced by α inside $\mathcal{R}^1_n(\mathcal{S})$, contains no minimal elements. The existence of such α follows from Proposition 5.2 and Lemma 3.1.

Fix a maximal c.e. set M and an element A from $\mathcal{S}$. Assume that $\overline{M} := \omega \setminus M = \{m_0 <_\omega m_1 <_\omega m_2 <_\omega \cdots\}$. Define

$$\beta(k) := \begin{cases} \alpha(e), & \text{if } k = m_e \text{ for some } e \in \omega, \\ A, & \text{if } k \in M. \end{cases}$$

Clearly, β is a E^1_n-computable numbering of $\mathcal{S}$. We claim that β is a *minimal* numbering of $\mathcal{S}$, i.e. β induces a minimal element inside $\mathcal{R}^1_n(\mathcal{S})$. Indeed, suppose that ξ is a numbering of $\mathcal{S}$, and a computable function f reduces

ξ to β. Then the maximality of M implies that the set $range(f) \setminus \overline{M}$ must be finite. This allows us to build a function g, which will reduce β to ξ, as follows:

(a) If $k \in range(f) \setminus \overline{M}$, then one can choose an appropriate value $g(k)$ in a non-uniform way.
(b) If $k \in M \cup range(f)$, then the image $g(k)$ will be defined either as some $l \in f^{-1}(k)$, or as a fixed a with $\xi(a) = A$.

A formal construction of the desired g can be recovered from Case 2 in Proposition 4.3, or from Theorem 1.3 in Ref.[5]. Recall that the principal ideal $\mathcal{I}$, induced by α, has no minimal elements. Hence, $\beta \not\leq \alpha$.

Fix different elements B and C from $\mathcal{S}$ such that $B \neq A \neq C$. Define

$$\gamma(k) := \begin{cases} B, & \text{if } k \in \emptyset', \\ C, & \text{if } k \notin \emptyset'. \end{cases}$$

Note that γ indexes *only* the finite family $\{B, C\}$. Thus, clearly, $\alpha \not\leq \gamma$ and $\beta \not\leq \gamma$.

Claim 5.1. *γ is not reducible to β.*

Proof. Assume that a function f reduces γ to β. Then $range(f) \subseteq \overline{M}$. Since the family $\mathcal{S}$ is infinite and M is maximal, we deduce that the set $range(f)$ must be finite, but this contradicts the non-computability of the set $\emptyset'$. □

From now on, we will employ the following useful fact without explicitly referencing it:

Lemma 5.1 (essentially Proposition 3.1 in Ref.[6]). *Let ζ, ξ_0, ξ_1 be arbitrary numberings. If $\zeta \leq \xi_0 \oplus \xi_1$, then at least one of the following conditions holds:*

(1) $\zeta \leq \xi_0$.
(2) $\zeta \leq \xi_1$.
(3) There are numberings ζ_0 and ζ_1 such that $\zeta_0 \leq \xi_0$, $\zeta_1 \leq \xi_1$, and $\zeta \equiv \zeta_0 \oplus \zeta_1$. Moreover, if the numberings ζ, ξ_0, and ξ_1 are E^1_n-computable, then both ζ_0 and ζ_1 are also E^1_n-computable.

Note that a similar fact has been already used in the proof of Proposition 4.1.

We define the desired E^1_n-computable numberings of $\mathcal{S}$:

$$\nu_0 := \alpha \oplus \gamma, \quad \nu_1 := \beta, \quad \mu := \gamma \oplus \beta.$$

Clearly, $\mu \leq \nu_0 \oplus \nu_1$.

Claim 5.2. *$\mu \not\leq \nu_0$ and $\mu \not\leq \nu_1$.*

Proof. Since $\gamma \not\leq \beta$, we deduce that $\mu \not\leq \nu_1$. Towards a contradiction, assume that $\gamma \oplus \beta \leq \alpha \oplus \gamma$. Then $\beta \leq \alpha \oplus \gamma$. Since $\beta \not\leq \alpha$ and $\beta \not\leq \gamma$, there are numberings β_0 and β_1 with $\beta \equiv \beta_0 \oplus \beta_1$, $\beta_0 \leq \alpha$, and $\beta_1 \leq \gamma$.

Clearly, any set $X \in \mathcal{S} \setminus \{B, C\}$ has a β_0-index. Moreover, by putting

$$\widetilde{\beta}_0(k) := \begin{cases} B, & \text{if } k = 0, \\ C, & \text{if } k = 1, \\ \beta_0(k-2), & \text{if } k \geq 2, \end{cases}$$

we obtain that the numbering $\widetilde{\beta}_0$ indexes *the whole* family $\mathcal{S}$, $\beta \equiv \widetilde{\beta}_0 \oplus \beta_1$, and $\widetilde{\beta}_0 \leq \alpha$. The minimality of β implies that $\widetilde{\beta}_0 \equiv \beta$. Hence, $\beta \leq \alpha$, which gives a contradiction. □

Assume, towards a contradiction, that (the degrees of) μ, ν_0, and ν_1 satisfy the weak distributivity property. Then there are E^1_n-computable numberings μ_0 and μ_1 of $\mathcal{S}$ such that $\mu \equiv \mu_0 \oplus \mu_1$, $\mu_0 \leq \nu_0$, and $\mu_1 \leq \nu_1$. Since $\nu_1 = \beta$ is minimal, we have $\mu_1 \equiv \beta$.

Clearly, $\mu_0 \not\leq \gamma$. Define a new numbering α_0 of the family $\mathcal{S}$ as follows:

(a) If $\mu_0 \leq \alpha$, then set $\alpha_0 := \mu_0$.
(b) Otherwise, there are numberings α^* and γ^* such that $\mu_0 \equiv \alpha^* \oplus \gamma^*$, $\alpha^* \leq \alpha$, and $\gamma^* \leq \gamma$. Put

$$\alpha_0(k) := \begin{cases} B, & \text{if } k = 0, \\ C, & \text{if } k = 1, \\ \alpha^*(k-2), & \text{if } k \geq 2. \end{cases}$$

Clearly, in each of the cases (a) and (b), α_0 is reducible to both α and μ_0.

Recall that $\mu_0 \leq \mu = \gamma \oplus \beta$ and hence, $\alpha_0 \leq \gamma \oplus \beta$. Obviously, $\alpha_0 \not\leq \gamma$. We define a numbering α_1 of $\mathcal{S}$:

(a) If $\alpha_0 \leq \beta$, then set $\alpha_1 := \alpha_0$.

(b) Otherwise, there are numberings γ' and β' such that $\alpha_0 \equiv \gamma' \oplus \beta'$, $\gamma' \leq \gamma$, and $\beta' \leq \beta$. Set

$$\alpha_1(k) := \begin{cases} B, & \text{if } k = 0, \\ C, & \text{if } k = 1, \\ \beta'(k-2), & \text{if } k \geq 2. \end{cases}$$

Clearly, we have $\alpha_1 \leq \alpha_0$ and $\alpha_1 \leq \beta$.

Since β is minimal, we deduce that $\beta \equiv \alpha_1 \leq \alpha_0 \leq \alpha$, which contradicts the original choice of the numbering α. Therefore, the numberings μ, ν_0, ν_1 witness the failure of weak distributivity. Theorem 5.1 is proved. □

6. Further discussion

After all the results of previous sections, it is completely possible that an interested reader would ask the following natural question:

Problem 6.1. *Let Γ be a class of the analytical hierarchy. What results on Rogers semilattices of Γ-computable families can be obtained, if one replaces **PD** with another set-theoretic assumption?*

Here we give a (very brief) case study for this problem: We assume the *Axiom of Constructibility* ($V = L$) and list some of results, which can be obtained under this assumption.

The Axiom of Constructibility says that every set is constructible. A formal statement of the axiom can be found, e.g., in Chap. 13 of Ref.[25].

Recall that the key property of a class E^1_n, which was heavily employed in the previous sections, is the prewellordering property (see Sec. 3.2.1).

Theorem 6.1 (see Exercises 5A.3 and 4B.10 of Ref.[34]). *Assume ($V = L$). For every $n \geq 3$, the class Σ^1_n has the prewellordering property. Consequently, Σ^1_n satisfies the reduction principle.*

Therefore, one can repeat the proofs of Theorems 4.1 and 4.2 verbatim, and obtain the following:

Corollary 6.1 (V = L). *Let $n \geq 3$, and let $\mathcal{S}$ be a finite family of Σ^1_n sets.*

(1) The Rogers semilattice $\mathcal{R}_{\Sigma^1_n}(\mathcal{S})$ has the greatest element if and only if the family $\mathcal{S}$ contains a least element under $\subseteq$.

(2) If $\mathcal{R}_{\Sigma^1_n}(\mathcal{S})$ has no greatest element, then every element from $\mathcal{R}_{\Sigma^1_n}(\mathcal{S})$ has a minimal cover.

In particular, we observe the following simple fact, which is still interesting on its own:

Remark 6.1. Let $\mathcal{S}$ be a finite family of Σ^1_3 sets.

(a) If one assumes **PD**, then $\mathcal{R}_{\Sigma^1_3}(\mathcal{S})$ has a greatest element iff $\mathcal{S}$ contains a *greatest* element under $\subseteq$ (Corollary 4.2).
(b) If one assumes $(V = L)$, then $\mathcal{R}_{\Sigma^1_3}(\mathcal{S})$ has a greatest element iff $\mathcal{S}$ contains a *least* element under $\subseteq$.

We strongly conjecture that one can employ the techniques developed by Tanaka[41] to provide a complete solution of Problem 1.2 under **PD** and under $(V = L)$.

As a concluding remark, we note the following: It seems that all our proofs essentially employed only the properties inherent to Spector pointclasses (see Section 4C of Ref.[34]). Hence, we formulate the following:

Problem 6.2. *Develop the theory of Rogers semilattices for Spector pointclasses.*

Acknowledgements

The work was supported by Nazarbayev University Faculty Development Competitive Research Grants 021220FD3851. The work of N. Bazhenov was carried out within the framework of the state contract of Sobolev Institute of Mathematics (project no. 0314-2019-0002). Part of the research contained in this paper was carried out while N. Bazhenov was visiting the Department of Mathematics of Nazarbayev University, Nur-Sultan. The authors wish to thank Nazarbayev University for its hospitality.

References

1. J. W. Addison. Separation principles in the hierarchies of classical and effective descriptive set theory. *Fundam. Math.*, 46(2):123–135, 1959.
2. J. W. Addison and Y. N. Moschovakis. Some consequences of the axiom of definable determinateness. *Proc. Natl. Acad. Sci. USA*, 59(3):708–712, 1968.
3. S. Badaev and S. Goncharov. Computability and numberings. In S. B. Cooper, B. Löwe, and A. Sorbi, editors, *New Computational Paradigms*, pages 19–34. Springer, New York, 2008.
4. S. Badaev, S. Goncharov, S. Podzorov, and A. Sorbi. Algebraic properties of Rogers semilattices of arithmetical numberings. In S. S. Goncharov and S. B. Cooper, editors, *Computability and Models*, pages 45–78. Springer, New York, 2003.

5. S. Badaev, S. Goncharov, and A. Sorbi. Completeness and universality of arithmetical numberings. In S. S. Goncharov and S. B. Cooper, editors, *Computability and Models*, pages 11–44. Springer, New York, 2003.
6. S. Badaev, S. Goncharov, and A. Sorbi. Isomorphism types and theories of Rogers semilattices of arithmetical numberings. In S. S. Goncharov and S. B. Cooper, editors, *Computability and Models*, pages 79–92. Springer, New York, 2003.
7. S. A. Badaev and S. S. Goncharov. Theory of numberings: open problems. In P. Cholak, S. Lempp, M. Lerman, and R. Shore, editors, *Computability Theory and Its Applications*, volume 257 of *Contemp. Math.*, pages 23–38. American Mathematical Society, Providence, 2000.
8. S. A. Badaev, S. S. Goncharov, and A. Sorbi. Elementary theories for Rogers semilattices. *Algebra Logic*, 44(3):143–147, 2005.
9. S. A. Badaev, S. S. Goncharov, and A. Sorbi. Isomorphism types of Rogers semilattices for families from different levels of the arithmetical hierarchy. *Algebra Logic*, 45(6):361–370, 2006.
10. S. A. Badaev and S. Yu. Podzorov. Minimal coverings in the Rogers semilattices of Σ^0_n-computable numberings. *Sib. Math. J.*, 43(4):616–622, 2002.
11. N. Bazhenov, M. Mustafa, and M. Yamaleev. Elementary theories and hereditary undecidability for semilattices of numberings. *Arch. Math. Logic*, 58(3–4):485–500, 2019.
12. N. Bazhenov, S. Ospichev, and M. Yamaleev. Isomorphism types of Rogers semilattices in the analytical hierarchy. To appear in Lect. Notes Ser., Inst. Math. Sci., Natl. Univ. Singapore. arXiv:1912.05226.
13. M. V. Dorzhieva. Elimination of metarecursive in Owing's theorem. *Vestn. Novosib. Gos. Univ., Ser. Mat. Mekh. Inform.*, 14(1):35–43, 2014. In Russian.
14. M. V. Dorzhieva. Undecidability of elementary theory of Rogers semilattices in analytical hierarchy. *Sib. Elektron Mat. Izv.*, 13:148–153, 2016. In Russian.
15. M. V. Dorzhieva. A single-valued numbering for the family of all Σ^1_2-sets. *Sib. Zh. Chist. Prikl. Mat.*, 18(2):47–52, 2018. In Russian.
16. Yu. L. Ershov. *Theory of numberings*. Nauka, Moscow, 1977. In Russian.
17. Yu. L. Ershov. Theory of numberings. In E. R. Griffor, editor, *Handbook of Computability Theory*, volume 140 of *Stud. Logic Found. Math.*, pages 473–503. North-Holland, Amsterdam, 1999.
18. Yu. L. Ershov. Necessary isomorphism conditions for Rogers semilattices of finite partially ordered sets. *Algebra Logic*, 42(4):232–236, 2003.
19. Yu. L. Ershov and I. A. Lavrov. The upper semilattice $L(\gamma)$. *Algebra Logic*, 12(2):93–106, 1973.
20. Ju. L. Eršov. Theorie der Numerierungen I. *Z. Math. Logik Grundlagen Math.*, 19(19–25):289–388, 1973.
21. Ju. L. Eršov. Theorie der Numerierungen II. *Z. Math. Logik Grundlagen Math.*, 21(1):473–584, 1975.
22. Ju. L. Eršov. Theorie der Numerierungen III. *Z. Math. Logik Grundlagen Math.*, 23(19–24):289–371, 1977.
23. K. Gödel. Über formal unentscheidbare Sätze der Principia Mathematica und verwandter Systeme. I. *Monatsh. Math. Phys.*, 38(1):173–198, 1931.

24. S. S. Goncharov and A. Sorbi. Generalized computable numerations and nontrivial Rogers semilattices. *Algebra Logic*, 36(6):359–369, 1997.
25. T. Jech. *Set Theory. The Third Millenium Edition, revised and expanded.* Springer, Berlin, 2002.
26. I. Sh. Kalimullin, V. G. Puzarenko, and M. Kh. Faizrakhmanov. Positive presentations of families in relation to reducibility with respect to enumerability. *Algebra Logic*, 57(4):320–323, 2018.
27. I. Sh. Kalimullin, V. G. Puzarenko, and M. Kh. Faizrakhmanov. Partial decidable presentations in hyperarithmetic. *Sib. Math. J.*, 60(3):464–471, 2019.
28. A. B. Khutoretskii. On the cardinality of the upper semilattice of computable enumerations. *Algebra Logic*, 10(5):348–352, 1971.
29. S. C. Kleene. *Introduction to metamathematics.* Van Nostrand, New York, 1952.
30. S. C. Kleene. Hierarchies of number-theoretic predicates. *Bull. Amer. Math. Soc.*, 61(3):193–213, 1955.
31. A. N. Kolmogorov and V. A. Uspenskii. On the definition of an algorithm. *Uspehi Mat. Nauk*, 13(4):3–28, 1958. In Russian.
32. D. A. Martin. The axiom of determinateness and reduction principles in the analytical hierarchy. *Bull. Amer. Math. Soc.*, 74(4):687–689, 1968.
33. Y. N. Moschovakis. Uniformization in a playful universe. *Bull. Amer. Math. Soc.*, 77(5):731–736, 1971.
34. Y. N. Moschovakis. *Descriptive Set Theory. Second Edition.* American Mathematical Society, Providence, 2009.
35. S. Yu. Podzorov. Initial segments in Rogers semilattices of Σ^0_n-computable numberings. *Algebra Logic*, 42(2):121–129, 2003.
36. S. Yu. Podzorov. Arithmetical D-degrees. *Sib. Math. J.*, 49(6):1109–1123, 2008.
37. H. Rogers. Gödel numberings of partial recursive functions. *J. Symb. Logic*, 23(3):331–341, 1958.
38. H. Rogers. *Theory of recursive functions and effective computability.* McGraw-Hill, New York, 1967.
39. G. E. Sacks. *Higher recursion theory.* Springer, Berlin, 1990.
40. V. L. Selivanov. Two theorems on computable numberings. *Algebra Logic*, 15(4):297–306, 1976.
41. H. Tanaka. Recursion theory in analytical hierarchy. *Comment. Math. Univ. St. Pauli*, 27(2):113–132, 1978.
42. V. A. Uspenskii. Systems of denumerable sets and their enumeration. *Dokl. Akad. Nauk SSSR*, 105:1155–1158, 1958. In Russian.
43. V. V. V'yugin. On some examples of upper semilattices of computable enumerations. *Algebra Logic*, 12(5):287–296, 1973.

Some notes on the *wtt*-jump

Klaus Ambos-Spies
University of Heidelberg, Germany
E-mail: ambos@math.uni-heidelberg.de

Rod Downey*
Victoria University of Wellington, New Zealand
E-mail: rod.downey@msor.vuw.ac.nz

Martin Monath
University of Heidelberg, Germany
E-mail: martin.monath@posteo.de

1. Introduction

One of the fundamental operations in classical computability is the jump operator. For a set B, $B' = \{e \mid \Phi_e^B(e) \downarrow\}$ is the canonical Σ_1^B complete set; the halting set *relative* to B. This notion goes back to Turing[17], but it is first highlighted in Kleene and Post[13]. The interplay between the jump operator and the partial ordering of the Turing degrees has provided one of the main areas of work in classical and applied computability theory.

This paper studies a notion called the *wtt*-jump. This is an operator akin to the classical Turing jump but using *wtt*-operators in place of Turing ones. The *wtt*-jump has previously been studied in one form or another by Coles, Downey and LaForte[6], Downey and Greenberg[8] and Anderson and Csima[2]. Recently, the authors have shown that properties of the *wtt*-jump can be used to characterize precisely when a c.e. set can be computed from a maximal c.e. set with a computably bounded use[1]. Earlier, Barmpalias, Downey and Greenberg[4], building on work by Chisholm, Chubb, Harizanov, Hirschfeldt, Jockusch, McNicholl and Pingrey[5] and by Afshari,

*Downey is partially supported by Marsden Fund of New Zealand. The results of this paper were obtained whilst Downey was an Alexander von Humboldt Fellow at the University of Heidelberg.

Barmpalias, Cooper and Stephan[3], characterized when all c.e. members of a c.e. Turing degree are computable from a hypersimple set and from an initial segment of a scattered linear ordering via a bounded reduction; but that characterization used what are called totally ω-c.a. degrees, and concerned approximations to total functions, whereas our maximal set result involved a new hierarchy generated from *wtt*-computations. Thus we see the following. In the same way that tt-reducibility turned out to be a central unifying idea in algorithmic randomness (e.g. Downey-Hirschfeldt[9]), these studies, and others, show that rather than mere artifacts of definitions in classical computability theory, hierarchies related to strong reducibilities and bounded jump operators can give classification and unification of combinatorics in parts of computable mathematics. As a consequence, it seems we should better understand analogs of the core notions of classical computability for such hierarchies. This paper contributes to that program.

The earliest analog of a jump operator using only bounded reducibilities is the "mini-jump" hierarchy introduced by Ershov[10] as discussed in Odifreddi[15], Chapter XI.6. Ershov's hierarchy concerned a jump operator for the m-degrees involving the partial m-degrees. Also a bounded analog of the jump for tt-reductions was investigated by Gerla[11].

As in Downey and Greenberg[8], we will let $\widehat{\Phi}_e$ denote the e-th *wtt*-procedure which will have use φ_e. It can be obtained from a standard listing of all pairs $\{\langle \Psi_e, \psi_e\rangle \mid e \in \mathbb{N}\}$ consisting of a partial Turing procedure and a partial computable function, by allowing $\Psi_e^A(x)\downarrow$ only if $\Psi_e^A(x)$ halts with information used $\leq \psi_e(x)\downarrow$. Note that $\{\varphi_e\}_{e\geq 0}$ is an acceptable numbering of the partial computable functions. So we may assume that the halting set is based on this numbering, i.e., $\emptyset' = \{e : \varphi_e(e)\downarrow\}$. Moreover, for any uniformly computable *wtt*-functionals $\{\widehat{\Psi}_e\}_{e\geq 0}$ (i.e., $\Psi^A(\langle e, x\rangle) = \widehat{\Psi}_e^A(x)$ is a *wtt*-functional) there are computable functions f and g such that $\widehat{\Psi}_e^A(x) = \widehat{\Phi}_{f(e)}^A(x) = \widehat{\Phi}_{g(\langle e,x\rangle)}^A(z)$ for all oracle sets A and all numbers e, x, z (see[1] or[8] for details). In the following we will tacitly use these facts.

Definition 1.1. The *wtt*-jump or bounded jump of a set A is defined by

$$A^{\dagger} = \{e \mid \widehat{\Phi}_e^A(e)\downarrow\}.$$

Clearly the usual equivalences obtained by the s-m-n theorem apply. So the *wtt*-jump of A is (up to m-degree) the same as $\{\langle e, n\rangle \mid \widehat{\Phi}_e^A(n)\downarrow\}$.

Note that $\emptyset' \equiv_m \emptyset^{\dagger}$, and that for a c.e. set A, $\emptyset' \leq_{wtt} A^{\dagger} \leq_{wtt} (\emptyset')^{\dagger}$. Moreover if X is Δ_2^0, $X^{\dagger}$ is also Δ_2^0.

The analog of the idea of lowness for the bounded jump can be defined as follows. A set A is *bounded low* if $A^\dagger \leq_{wtt} \emptyset'$ (or, equivalently, $A^\dagger \leq_{tt} \emptyset'$). Variations of bounded lowness– all of them wtt-equivalent to this notion – have been studied by Anderson and Csima[2], Ambos-Spies, Downey and Monath[1], and Wu and Wu[18]. It is easy to see that all classically superlow sets A (i.e. $A' \leq_{wtt} \emptyset'$) are bounded low, but in Ambos-Spies, Downey and Monath[1], we proved that there are Turing complete bounded low c.e. sets. (This result was independently obtained by Wu and Wu[18].)

2. Arithmetical characterizations

Anderson and Csima[2] proved that if $X \geq_{wtt} \emptyset'$ then there is a set A with $A^\dagger \equiv_{wtt} X$. We will be concerned with an analog of Sacks Jump Theorem[16], and hence we will need an analog of the arithmetical hierarchy. The classical Sacks Jump Theorem[16] says that if X is Σ^0_2 then there is a c.e. set A with $A' \equiv_T \emptyset' \oplus X$.

The problem is that relativization is not so straightforward. What should Σ^0_2 mean in this setting? Here are some candidates:

Definition 2.1.

(1) A set X is $\widehat{\Sigma}^B_1$ iff there is a partial computable function g and a relation R which is g-bounded computable with

$$x \in X \text{ iff } \exists s R^B(x,s).$$

Here by R being *g-bounded computable* we mean that $R^B(x,s)$ holds requires that $g(x)$ is defined and any query in the computation of $R^B(x,s)$ is bounded by $g(x)$.

(2) If the function g above can be chosen to be *total* computable then we say that X is $\widetilde{\Sigma}^B_1$.

Correspondingly, we say that X is $\widehat{\Pi}^B_1$ iff $\overline{X}$ is $\widehat{\Sigma}^B_1$, and X is $\widehat{\Delta}^B_1$ iff $X \in \widehat{\Pi}^B_1 \cap \widehat{\Sigma}^B_1$. Similarly, X is $\widetilde{\Pi}^B_1$ iff $\overline{X}$ is $\widetilde{\Sigma}^B_1$, and X is $\widetilde{\Delta}^B_1$ iff $X \in \widetilde{\Pi}^B_1 \cap \widetilde{\Sigma}^B_1$. Obviously, $\widetilde{\Sigma}^B_1 \subseteq \widehat{\Sigma}^B_1$, $\widetilde{\Pi}^B_1 \subseteq \widehat{\Pi}^B_1$ and $\widetilde{\Delta}^B_1 \subseteq \widehat{\Delta}^B_1$. In the following proposition we give some properties of the bounded Σ_1-, Π_1- and Δ_1-classes. In particular we show that, for some oracle sets B, the latter inclusions are proper. On the other hand, note that, for the empty oracle, obviously $\widetilde{\Sigma}^\emptyset_1 = \widehat{\Sigma}^\emptyset_1 = \Sigma^\emptyset_1$ etc.

Proposition 2.1.

(i) $X \in \widehat{\Sigma}^B_1$ iff $X = dom(\widehat{\Phi}^B_e)$ for some $e \geq 0$, and $X \in \widetilde{\Sigma}^B_1$ iff $X = dom(\widehat{\Phi}^B_e)$ for some $e \geq 0$ such that φ_e is total.

(ii) $X \in \widehat{\Sigma}^B_1$ *iff* $X \leq_m B^\dagger$.
(iii) $X \in \widetilde{\Delta}^B_1$ *iff* $X \leq_{wtt} B$.
(iv) There is a c.e. set B *such that* $\widetilde{\Delta}^B_1 \subset \widehat{\Delta}^B_1$, $\widetilde{\Sigma}^B_1 \subset \widehat{\Sigma}^B_1$ *and* $\widetilde{\Pi}^B_1 \subset \widehat{\Pi}^B_1$.

Proof. (i) We prove the first part. The proof of the second part is similar.

First assume that $X \in \widehat{\Sigma}^B_1$. Fix a partial computable function g and a g-bounded computable relation R such that $x \in X$ iff $R^B(x,s)$ for some s. Then X is the domain of the *wtt*-functional Ψ where $\Psi^B(x)$ is defined iff there is a number s such that $R^B(x,s)$ holds (note that the use of Ψ is bounded by g). So, for any e, such that $\Psi = \widehat{\Phi}_e$, $X = dom(\widehat{\Phi}^B_e)$.

For the converse direction, assume that $X = dom(\widehat{\Phi}^B_e)$. Fix R such that $R^B(x,s)$ holds iff $\widehat{\Phi}^B_{e,s}(x)$ is defined. Then R is g-bounded computable for the partial computable function $g = \varphi_e$ and $x \in X$ iff $R^B(x,s)$ holds for some s. So X is $\widehat{\Sigma}^B_1$.

(ii) First assume that $X \in \widehat{\Sigma}^B_1$. By (i) fix e such that $X = dom(\widehat{\Phi}^B_e)$. Then, for a total computable function f such that $\widehat{\Phi}^B_{f(x)}(z) = \widehat{\Phi}^B_e(x)$ for all x and z, $X \leq_m B^\dagger$ via f.

For the converse direction, assume that $X \leq_m B^\dagger$ via f. Then $x \in X$ iff $\widehat{\Phi}^B_{f(x)}(f(x))$ is defined. So, for R such that $R^B(x,s)$ holds iff $\widehat{\Phi}^B_{f(x),s}(f(x))$ is defined and for $g(x) = \varphi_{f(x)}(f(x))$, g is partial computable, R is g-bounded computable, and $x \in X$ iff $R^B(x,s)$ holds for some s. So X is $\widehat{\Sigma}^B_1$.

(iii) First assume that $X \in \widetilde{\Delta}^B_1$, i.e., $X \in \widetilde{\Sigma}^B_1$ and $\overline{X} \in \widetilde{\Sigma}^B_1$. Fix total computable functions g_0 and g_1 and relations R_0 and R_1 such that R_0 is g_0-bounded computable and $x \in X$ iff $R^B_0(x,s)$ for some s and R_1 is g_1-bounded computable and $x \in \overline{X}$ iff $R^B_1(x,s)$ for some s. Then $X \leq_{wtt} B$ via the *wtt*-functional $\widehat{\Psi}$ where $\widehat{\Psi}^B(x) = 1$ if for the least s such that $R^B_0(x,s)$ or $R^B_1(x,s)$ holds, $R^B_0(x,s)$ holds, and $\widehat{\Psi}^B(x) = 0$ otherwise (note that $\widehat{\Psi}$ is g-bounded for $g(x) = \max\{g_0(x), g_1(x)\}$).

For the converse direction, assume that $X \leq_{wtt} B$, say $X = \widehat{\Phi}^B_e$ where φ_e is total. Then, for R_0 and R_1 such that $R_0(x,s)$ holds if $\widehat{\Phi}^B_{e,s}(x) = 1$ and $R_1(x,s)$ holds if $\widehat{\Phi}^B_{e,s}(x) = 0$ and for $g = \varphi_e$, g is total computable, R_0 and R_1 are g-bounded computable, $x \in X$ iff $R_0(x,s)$ holds for some s, and $x \in \overline{X}$ iff $R_1(x,s)$ holds for some s. So $X \in \widetilde{\Sigma}^B_1$ and $\overline{X} \in \widetilde{\Sigma}^B_1$ hence $X \in \widetilde{\Delta}^B_1$.

(iv) By Downey and Greenberg[8], Proposition 3.1, there is a c.e. set B and a set X such that $X \in \widehat{\Delta}^B_1$ and $X \not\leq_{wtt} B$. So, by (iii), $X \in \widehat{\Delta}^B_1 \setminus \widetilde{\Delta}^B_1$. This implies the first part of the claim. Moreover, by definition, $X \in \widehat{\Sigma}^B_1$ and $\overline{X} \in \widehat{\Sigma}^B_1$ (hence $\overline{X} \in \widehat{\Pi}^B_1$ and $X \in \widehat{\Pi}^B_1$) whereas $X \notin \widetilde{\Sigma}^B_1$ or $\overline{X} \notin \widetilde{\Sigma}^B_1$ (hence $\overline{X} \notin \widetilde{\Pi}^B_1$ or $X \notin \widetilde{\Pi}^B_1$). Obviously this implies the second and third parts of the claim. □

The most natural guess for an analog of Σ_1^B is $\widehat{\Sigma}_1^B$. By Proposition 2.1 (i) the sets in this class are just the sets which are bounded-c.e. in B in the sense of[8]. As (ii) shows, the characterization of the classical Σ_1^B classes as the classes of the sets which are many-one reducible to the jump of B carries over to $\widehat{\Sigma}_1^B$ if we replace the jump by the bounded jump. But (iii) together with (iv) shows a serious drawback to this definition: for some sets B, there are $\widehat{\Delta}_1^B$ sets which are not bounded computable from B. The more limited definition of $\widetilde{\Sigma}_1^B$ does not share this problem, but here the analog of (ii) fails. In particular, since $\widetilde{\Sigma}_1^B$ is downward closed under many-one reducibility, there is a set B such that the bounded jump $B^\dagger$ is not in $\widetilde{\Sigma}_1^B$ (by (ii) and (iv)).

Exactly as in the arithmetical hierarchy, we can now define $\widehat{\Sigma}_n^B$ and $\widetilde{\Sigma}_n^B$ for all $n \geq 1$. For example, X is $\widehat{\Sigma}_2^B$ means that there is a $\widehat{\Pi}_1^B$ set Q such that X is $\widehat{\Sigma}_1^Q$. If $B = \emptyset$ we say that X is $\widehat{\Sigma}_2^0$.

The following proposition is a straightforward analog of the corresponding fact in classical computability theory.

Proposition 2.2.

(i) X is $\widehat{\Sigma}_2^0$ iff X is $\widehat{\Sigma}_1^{\emptyset'}$.
(ii) X is $\widetilde{\Sigma}_2^0$ iff X is $\widetilde{\Sigma}_1^{\emptyset'}$.

Proof. For the following recall that the classes $\widehat{\Sigma}_1^0$ and $\widetilde{\Sigma}_1^0$ coincide with the class of the c.e. sets.

(i) First suppose that X is $\widehat{\Sigma}_2^0$. Then there is a $\widehat{\Pi}_1^0$ set Q with $X \in \widehat{\Sigma}_1^Q$. Since $\widehat{\Sigma}_1^{\overline{Q}} = \widehat{\Sigma}_1^Q$ there is a partial computable function g and a g-bounded computable relation R such that

$$x \in X \text{ iff } \exists s R^{\overline{Q}}(x, s).$$

Since $\overline{Q}$ is in $\widehat{\Sigma}_1^0$ hence c.e., $\overline{Q} \leq_m \emptyset'$; say via h. So any g-bounded query to $\overline{Q}$ can be replaced by an $h(g)$-bounded query to $\emptyset'$. So there is an $h(g)$-bounded computable relation $\hat{R}$ such that

$$x \in X \text{ iff } \exists s \hat{R}^{\emptyset'}(x, s).$$

Hence X is $\widehat{\Sigma}_1^{\emptyset'}$.

For the opposite direction, suppose that X is $\widehat{\Sigma}_1^{\emptyset'}$. Then X is $\widehat{\Sigma}_1^{\overline{\emptyset'}}$ too. Since $\emptyset'$ is c.e. and the class of the c.e. sets coincides with $\widehat{\Sigma}_1^0$, it follows that $\overline{\emptyset'}$ is in $\widehat{\Pi}_1^0$. So X is $\widehat{\Sigma}_2^0$ by definition.

(ii) The proof is similar to the proof of part (i). For the first direction it suffices to note that if g is total then, by totality of h, $h(g)$ is total too. For the second direction, it suffices to use the observation that $\widetilde{\Pi}^0_1 = \widehat{\Pi}^0_1$. Hence $\overline{\emptyset'} \in \widetilde{\Pi}^0_1$. □

3. Potential analogs of the Sacks Jump Theorem

We have pointed out that *wtt*-superlow sets are useful in classifying the internal combinatorics of constructions. In this section, we will explore how the *wtt*-jump operator operates as an analog of the normal jump operator. As mentioned above, one such result was by Anderson and Csima[2] who proved that the natural analog of the Friedberg Jump Theorem holds. There have been earlier studies looking at the relationship between the classical jump and strong reducibilities, such as Mohrherr[14] and Csima, Downey and Ng[7]. Solving an open question raised in[2] asking for an analog of Sacks Jump Theorem for the *wtt*-jump, by examining bounded versions of the Sacks Jump Theorem[16]. Note that, by $\emptyset' \equiv_m \emptyset^\dagger$, we may replace $\emptyset'$ by $\emptyset^\dagger$ in the following results.

For any c.e. set A, $A^\dagger \in \widehat{\Sigma}^0_2$ (namely $A^\dagger \in \widehat{\Sigma}^A_1$ hence, by $A \leq_{wtt} \emptyset'$ and by Proposition 2.2, $A^\dagger \in \widehat{\Sigma}^{\emptyset'}_1 = \widehat{\Sigma}^0_2$). So, by $\emptyset' \leq_m A^\dagger$, $A^\dagger \equiv_{wtt} \emptyset' \oplus S$ for the $\widehat{\Sigma}^0_2$-set $S = A^\dagger$. This suggest that the full analog of the Sacks Jump Theorem for the bounded jump should be the statement that, for any $\widehat{\Sigma}^0_2$-set S, there is a c.e. set A such that $A^\dagger \equiv_{wtt} \emptyset' \oplus S$. As the following theorem shows, this strong analog fails.

Theorem 3.1. *There exists $S \in \widehat{\Sigma}^0_2$ such that, for no c.e. set W, $W^\dagger \equiv_{wtt} \emptyset' \oplus S$.*

Proof. The argument is finite injury. We will build a $\widehat{\Sigma}^0_2$ set S to meet the requirements

$$R_e : \neg(\Delta_e^{W_e^\dagger} = \emptyset' \oplus S \;\&\; \Gamma_e^{\emptyset' \oplus S} = W_e^\dagger).$$

Here, in the context of this proof, $\{(W_e, \Delta_e, \Gamma_e)\}_{e \geq 0}$ is an enumeration of all triples of a c.e. set and two *wtt*-functionals. Moreover, we assume that δ_e and γ_e are partial computable bounds on the uses of Δ_e and Γ_e, respectively, and we fix computable enumerations $W_{e,s}$ etc. of W_e etc. Moreover, we let $\delta^*_e(n) = \max_{n' \leq n} \delta_e(n')$ and $\gamma^*_e(n) = \max_{n' \leq n} \gamma_e(n')$, (where $\delta^*_e(n) \uparrow$ if $\delta_e(n') \uparrow$ for some $n' \leq n$, and similarly for $\gamma^*_e(n)$) and, w.l.o.g. we assume that $\delta_e(n) \geq n$ if defined, and similarly for $\gamma_e(n)$.

Before we can describe our strategy for meeting the requirements, we have to present some of the fundamentals of the construction first.

The format of the definition of S is as follows. Besides a computable approximation $\{S_s\}_{s\geq 0}$ of S we give a computable enumeration $\{B_s\}_{s\geq 0}$ of a c.e. set B and define a partial computable function ψ such that the approximation $\{S_s\}_{s\geq 0}$ of S is ψ-bounded controlled by B via $\{B_s\}_{s\geq 0}$ in the following sense. For any number z, $S_0(z) = 0$, $\psi(z)$ is defined if $S_t(z) = 1$ for some stage t, and $B_{s+1} \upharpoonright \psi(z) \neq B_s \upharpoonright \psi(z)$ for any s such that z is extracted from S at stage $s+1$, i.e., $z \in S_s \setminus S_{s+1}$. Obviously this ensures that $S = dom(\Psi^B)$ for some ψ-bounded wtt-functional Ψ whence $S \in \widehat{\Sigma}_1^B$. Since $\widehat{\Sigma}_1^{\overline{B}} = \widehat{\Sigma}_1^B$ and $\overline{B} \in \Pi_1^0 = \widehat{\Pi}_1^0$, this guarantees that S is in $\widehat{\Sigma}_2^0$.

In addition, we define uniformly computable approximations $\{C_{e,s}\}_{s\geq 0}$ of auxiliary sets C_e together with uniformly partial computable functions ξ_e such that $\{C_{e,s}\}_{s\geq 0}$ is ξ_e-bounded controlled by W_e via $\{W_{e,s}\}_{s\geq 0}$ (in the above sense). So there are uniformly computable ξ_e-bounded wtt-functionals Ξ_e such that $C_e = dom(\Xi_e^{W_e})$. It follows that there are total uniformly computable functions f_e such that $C_e(x) = W_e^\dagger(f_e(x))$, and, by the Recursion Theorem, we may assume that the functions f_e are given in advance.

We work with the standard approximation $\{W_{e,s}^\dagger\}_{s\geq 0}$ of $W_e^\dagger$ where $W_{e,s}^\dagger(m) = 1$ iff $\widehat{\Phi}_{m,s}^{W_{e,s}}(m)\downarrow$. So, if m is extracted from $W_e^\dagger$ at stage $s+1$, i.e., $m \in W_{e,s}^\dagger \setminus W_{e,s+1}^\dagger$, then $\varphi_m(m)$ is defined at stage s and a number $< \varphi_m(m)$ is enumerated into W_e at stage $s+1$, i.e., $\varphi_{m,s}(m)\downarrow$ and $W_{e,s+1} \upharpoonright \varphi_m(m) \neq W_{e,s} \upharpoonright \varphi_m(m)$. Finally, recall that we assume that any given functional defined at stage s has use $< s$ and any given partial computable function defined at stage s has value $< s$.

Now, for a single requirement R_e, we work as follows. For readability we will drop the index "e" in the following.

In order to monitor the "controlled" length of agreement of the equations in R_e at stage s we define the length functions

$$l_\Delta(s) = \max\{n : \forall n' < n(\Delta_s^{W_s^\dagger}(n') = (\emptyset'_s \oplus S_s)(n') \;\&\; \Gamma_s^{\emptyset'_s \oplus S_s} \upharpoonright \delta(n') = W_s^\dagger \upharpoonright \delta(n'))\}$$

and

$$l_\Gamma(s) = \max\{n : \forall n' < n(\Gamma_s^{\emptyset'_s \oplus S_s}(n') = W_s^\dagger(n') \;\&\; \Delta_s^{W_s^\dagger} \upharpoonright \gamma(n') = (\emptyset'_s \oplus S_s) \upharpoonright \gamma(n'))\},$$

let

$$l(s) = \min\{l_\Delta(s), l_\Gamma(s)\},$$

and call a stage s *expansionary* if $l(s) > l(t)$ for all stages $t < s$. Note that the limit of l exists (since the functionals involved in the definition are bounded) and the limit is finite iff there are only finitely many expansionary stages iff the requirement R_e is met.

We will work with a number z and a finite sequence of numbers x_i under our control, where z is targeted for S while the numbers x_i are targeted for C (hence $f(x_i)$ is targeted for $W^\dagger$ implicitly). The numbers z and $f(x_i)$ are potential diagonalization witnesses for the first equation and the second equation, respectively, in R_e. In the first phase of the attack we define the set up to be used in the second phase of the attack.

First pick an unused number z (above the restraint of the higher priority requirements) and wait for an expansionary stage s such that $l(s) > 2z+1$ (note that $S(z) = (\emptyset' \oplus S)(2z+1)$). Then $\delta^*(2z+1)$ is defined at stage s and we may pick $\delta^*(2z+1)$ unused numbers x_i such that $z < x_1 < \cdots < x_{\delta^*(2z+1)}$. Finally, wait for the least expansionary stage $s_0 > s+1$ (hence $s_0 \geq \delta^*(2z+1)+2$ and $l(s_0) > 2z+1$) such that, for $1 \leq n \leq \delta^*(2z+1)$, $\gamma^*(f(x_n))$ is defined at stage s_0, $s_0 > \gamma^*(f(x_n))$, and $l(s_0) > f(x_n)$. Then let $\psi(z) = s_0^3$ thereby completing the set up of the attack. (Note that if the set up cannot be completed then there are only finitely many expansionary stages. So R_e is met.)

Now in the second phase of the attack, a state is assigned to any stage $s \geq s_0$. Here the state of s is the number of $m < \delta^*(2z+1)$ such that $\varphi_m(m)$ is defined at stage s but is not yet defined at stage s_0, i.e.,

$$|\{m < \delta^*(2z+1) : \varphi_{m,s}(m)\downarrow \ \text{ and } \ \varphi_{m,s_0}(m)\uparrow\}|,$$

and we let s_n be the least expansionary stage $s \geq s_0$ of state n (if any). Note that any state is $\leq \delta^*(2z+1) < s_0 - 1$ and the state is nondecreasing in the stage. So, eventually, all stages have the same state, called the final state. Now, for any state n we have a procedure which is active at the stages of state n, i.e., it is started at stage $s_n + 1$ (if s_n is defined) and it becomes abandoned at stage s_{n+1} (if defined). Each n-procedure may change the approximative value of $S(z)$, but we guarantee that for each n the number of changes is bounded by s_0^2. Since the number of states is bounded by s_0, it follows that the total number of $S(z)$-changes is bounded by $\psi(z) = s_0^3$. So we may witness any extraction of z out of S by a change of B below $\psi(z)$. For $n \geq 1$, the n-procedure may also change the value of $C(x_n)$ hence $W^\dagger(f(x_n))$. Again we will guarantee that these changes are conform with making C uniformly bounded c.e. in W. We will show that a procedure which is never abandoned will guarantee that there are only finitely many expansionary stages. Since, for the final state n, the n-procedure is never abandoned, this shows that our strategy meets requirement R_e.

The 0*-procedure*. For any expansionary stage $t \geq s_0$, let $S_{t+1}(z) \neq S_t(z)$. If $z \in S_t$ then enumerate the least $y < \psi(z)$ which is not yet in B into B_{t+1} (we will argue below that such a y must exist).

Note that, for any two consecutive expansionary stages $t < t'$ of state 0, this ensures that $W^\dagger_{t'} \upharpoonright \delta^*(2z+1) \neq W^\dagger_t \upharpoonright \delta^*(2z+1)$. Since there are $\delta^*(2z+1)$ numbers m less than $\delta^*(2z+1)$ and since $W^\dagger_s(m)$ is 0-1-valued, for any set of $\delta^*(2z+1)+2$ consecutive such stages there must be at least one pair of consecutive expansionary stages $t < t'$ in the set such that $m \in W^\dagger_t \setminus W^\dagger_{t'}$ for some number $m < \delta^*(2z+1)$. So $W_{t'} \upharpoonright \varphi_m(m) \neq W_t \upharpoonright \varphi_m(m)$. But since t' has state 0, $\varphi_m(m)$ is defined at stage s_0 hence less than s_0. So $W_{t'} \upharpoonright s_0 \neq W_t \upharpoonright s_0$. It follows that, if the 0-procedure is never abandoned then the number of expansionary stages $\geq s_0$ is bounded by

$$\#_0 = (\delta^*(2z+1)+2) \cdot s_0 \leq s_0^2,$$

hence R_e is met. Moreover, the 0-procedure enumerates at most $\#_0$ many numbers (hence at most s_0^2 many numbers) $\leq \psi(z)$ into B.

The n-procedure $(n \geq 1)$. At stage $s_n + 1$ (if s_n exists) let $\xi(x_n) = s_n$ be the W-use of x_n and proceed as follows (note that $x_n \notin C_{s_n}$).

(*) Wait for the next expansionary stage t such that $x_n \notin C_t$ and $C_t(x_n) = W^\dagger_t(f(x_n))$. Put x_n into C at stage $t+1$, and go to (**).

(**) At the first expansionary stage $t' > t$ such that $C_{t'}(x_n) = W^\dagger_{t'}(f(x_n))$ let $S_{t'+1}(z) \neq S_{t'}(z)$. If $z \in S_{t'}$ then enumerate the least $y < \psi(z)$ which is not yet in B into $B_{t'+1}$ (we will argue below that such a y must exist).
At the first stage $t'' > t'$ such that $W_{t''+1} \upharpoonright s_n \neq W_{t''} \upharpoonright s_n$ (if any) remove x_n from C at stage $t''+1$, i.e., let $C_{t''+1}(x_n) = 0$, and return to (*). While waiting for such a stage t'', at any expansionary stage $u > t'$ let $S_{u+1}(z) \neq S_u(z)$. If $z \in S_u$ then enumerate the least $y < \psi(z)$ which is not yet in B into B_{u+1} (we will argue below that such a y must exist).

The success of the n-procedure, provided that it is never abandoned, and its conformity with the required features of the definitions of $C(x_n)$ and $S(z)$ are shown as follows. First note that x_n is extracted from C only at stages $t''+1$ such that $W_{t''+1} \upharpoonright s_n \neq W_{t''} \upharpoonright s_n$. Since $\xi(x_n) = s_n$, this shows that the approximation of $C(x_n)$ is as required. This justifies the assumption that $C(x_n) = W^\dagger(f(x_n))$, and we may conclude that $C_s(x_n) = W^\dagger_s(f(x_n))$ for all sufficiently large stages s. Next observe that, for any stage t as in (*) and the corresponding stage $t' > t$ in (**), $C_t(x_n) = W^\dagger_t(f(x_n)) =$

0 whereas $C_{t'}(x_n) = W^{\dagger}_{t'}(f(x_n)) = 1$. Moreover, since the stages t and t' are expansionary and greater than s_0, $W^{\dagger}_{t}(f(x_n)) = \Gamma_t^{\emptyset'_t \oplus S_t}(f(x_n))$ and $W^{\dagger}_{t'}(f(x_n)) = \Gamma_{t'}^{\emptyset'_{t'} \oplus S_{t'}}(f(x_n))$. Since S is not changed at any stage $\hat{t}$ with $t < \hat{t} \leq t'$, this implies that $\emptyset'_{t'} \restriction \gamma^*(f(x_n)) \neq \emptyset'_t \restriction \gamma^*(f(x_n))$. So, by $\gamma^*(f(x_n)) < s_0$, we may conclude that the procedure (*) is called $< s_0$ times. Since (*) and (**) are run alternatingly, (**) is called $< s_0$ times, too. Moreover, for any such call, we will change the approximation of $S(z)$ at most $\delta^*(2z+1)+2$ times. Namely, just as in the analysis of the 0-procedure, we may argue that if we change $S(z)$ so many times then one of theses changes must result in a change of W below $\varphi_m(m)$ for a number m such that $\varphi_m(m)$ is defined at stage s_n hence is less than s_n. So we return to (*) at this stage. It follows that the total number of $S(z)$-changes which may be caused by the n-procedure is bounded by $s_0 \cdot (\delta^*(2z+1)+2) < s_0^2$, hence the required permittings by B will be provided. Finally, if the n-attack is not abandoned, then R_e is met since the above bounds on the calls of the subprocedures (*) and (**) and the bounds on the $S(z)$-changes imply that there are only finitely many expansionary stages.

This completes the description of the strategy to meet a single requirement R_e and the proof of its correctness. This result follows by a standard application of the finite injury method. We leave this to the reader. □

Given the above failure of the potential strong analog of the Sacks Jump Theorem for the bounded jump, we may look for some weaker analogs. The following seems a good candidate, but this is illusory.

Proposition 3.1 (Tilde Sacks Jump Theorem). *Let S be $\widetilde{\Sigma}^0_2$. There is a c.e. set A with $A^{\dagger} \equiv_{wtt} \emptyset' \oplus S$.*

Originally we had a direct proof, but subsequently realized that the "tilde" version of the jump theorem is of limited interest as the jump hierarchy collapses, and Proposition 3.1 follows easily from the material below.

Proposition 3.2.

(i) For any set X, X is in $\widetilde{\Sigma}^0_2$ iff X is in $\widetilde{\Sigma}^{\emptyset'}_1$ iff X is in $\widetilde{\Delta}^{\emptyset'}_1$ iff $X \leq_{wtt} \emptyset'$.
(ii) For any $n \geq 2$, $\widetilde{\Sigma}^0_n = \widetilde{\Pi}^0_n = \widetilde{\Delta}^0_n = \widetilde{\Delta}^0_2 = \{X : X \leq_{wtt} \emptyset'\}$.

Proof. (i) By Proposition 2.2 (ii) and Proposition 2.1 (iii), it suffices to show that any set in $\widetilde{\Sigma}^{\emptyset'}_1$ is wtt-reducible to $\emptyset'$. So fix $X \in \widetilde{\Sigma}^{\emptyset'}_1$. By Proposition 2.1 (i) there is an index e such that $X = dom(\widehat{\Phi}^{\emptyset'}_e)$ where the bound φ_e on the use of $\widehat{\Phi}_e$ is total. Define the c.e. set D by

$$D = \{\langle x, \sigma \rangle : |\sigma| = \varphi_e(x) \ \&\ \widehat{\Phi}_e^{\sigma}(x)\downarrow\},$$

and fix a total computable function f such that $D \leq_m \emptyset'$ via f. Then, for any $x \geq 0$,

$$x \in X \Leftrightarrow \widehat{\Phi}_e^{\emptyset'}(x)\downarrow \Leftrightarrow \widehat{\Phi}_e^{\emptyset' \restriction \varphi_e(x)}(x)\downarrow \Leftrightarrow f(\langle x, \emptyset' \restriction \varphi_e(x)\rangle) \in \emptyset'.$$

Obviously, this implies $X \leq_{wtt} \emptyset'$ where the use of the reduction is bounded by the total computable function

$$g(x) = \max\{\varphi_e(x), 1 + \max_{|\sigma|=\varphi_e(x)} f(\langle x, \sigma \rangle)\}.$$

(ii) The proof is by induction on n. For $n = 2$ the claim is immediate by part (i) of the proposition since $X \leq_{wtt} \emptyset'$ iff $\overline{X} \leq_{wtt} \emptyset'$. So fix $n+1 > 2$ and assume (ii) for n. By symmetry and by (i), given $X \in \widetilde{\Sigma}^0_{n+1}$ it suffices to show that $X \leq_{wtt} \emptyset'$. Fix $Y \in \widetilde{\Pi}^0_n$ such that $X \in \widetilde{\Sigma}^Y_1$. Note that $\widetilde{\Sigma}^A_1 \subseteq \widetilde{\Sigma}^B_1$ for any sets A and B such that $A \leq_{wtt} B$. So, since $Y \leq_{wtt} \emptyset'$ by inductive hypothesis, it follows that X is in $\widetilde{\Sigma}^{\emptyset'}_1$. By (i) this implies $X \leq_{wtt} \emptyset'$. ☐

Note that Proposition 3.2 immediately implies Proposition 3.1 above if we let $A = \emptyset$. Namely, for $S \in \widetilde{\Sigma}^0_2$, it follows by Proposition 3.2 that $S \leq_{wtt} \emptyset'$ hence $A^\dagger = \emptyset^\dagger \equiv_{wtt} \emptyset' \equiv_{wtt} \emptyset' \oplus S$. This shows that we can hardly claim that Proposition 3.1 is an analog of the Sacks Jump Theorem for the bounded jump.

In order to get stronger and more satisfying analogs we have to consider natural extensions of the tilde classes. One possible extension to Definition 2.1 (ii), more based on the complexity of approximations akin to the hierarchies of Downey and Greenberg[8], would be again to take g to be total but to relax the requirement that g is computable by having g wtt-reducible to the oracle.

Definition 3.1. A set X is $\breve{\Sigma}^B_1$ iff there is a function g and a relation R such that $g \leq_{wtt} B$, R is g-bounded computable and

$$x \in X \text{ iff } \exists s R^B(x, s).$$

Unfortunately, this does little for us as (the first part of) the following shows.

Proposition 3.3.

(1) $\widehat{\Sigma}^{\emptyset'}_1 \subseteq \breve{\Sigma}^{\emptyset'}_1$.

(2) For all $X \in \breve{\Sigma}_1^{\emptyset'}$, there exists $Y \in \widehat{\Sigma}_1^{\emptyset'}$ such that $X \leq_{dtt} Y$ (meaning disjunctive truth-table reduction).
(3) $\widehat{\Sigma}_1^{\emptyset'}$ is a proper subclass of $\breve{\Sigma}_1^{\emptyset'}$.

Proof. For 1. Suppose that $X \in \widehat{\Sigma}_1^{\emptyset'}$. Fix a partial computable function g and a g-bounded computable relation R such that $x \in X$ iff $\exists s R^{\emptyset'}(x,s)$. Then, for h defined by $h(x) = 0$ if $g(x) \uparrow$ and $h(x) = g(x)$ if $g(x) \downarrow$, R is h-bounded computable. So it suffices to show that $h \leq_{wtt} \emptyset'$. But this is immediate by the s-m-n Theorem: for computable f such that $g(x) \downarrow$ iff $\varphi_{f(x)}(f(x)) \downarrow$ (i.e., $f(x) \in \emptyset'$), $h(x)$ can be computed from $\emptyset' \upharpoonright f(x)+1$.

For 2. Let $X \in \breve{\Sigma}_1^{\emptyset'}$ via $g(x)$ and $R^{\emptyset'}(x,s)$. By $g \leq_{wtt} \emptyset'$ we may fix a computable function h and a computable approximation $g(x) = \lim_s g(x,s)$ which changes at most $h(x) - 1$ many times. (Recall that $g \leq_{wtt} \emptyset'$ iff g is ω-c.a. by the proof of the Limit Lemma.) W.l.o.g. we may assume that $g(x,s)$ is nondecreasing in s (since we may replace $g(x,s)$ and $g(x)$ by $\hat{g}(x,s) = \max_{t \leq s} g(x,t)$ and $\hat{g}(x) = \lim_s \hat{g}(x,s)$, respectively). Similarly, we may assume that $R^{\emptyset'}(x,t)$ holds for all $t \geq s$ if $R^{\emptyset'}(x,s)$ holds.

Now, define the partial computable function ψ by letting $\psi(\langle x,i \rangle) = g(x,s)$ for the least s such that $|\{t < s : g(x,t) < g(x,t+1)\}| = i$ (if such an s exists, and $\psi(\langle x,i \rangle) \uparrow$ otherwise) and define the relation $\hat{R}$ by letting $\hat{R}^Z(\langle x,i \rangle, s)$ hold if $\psi(\langle x,i \rangle) \downarrow \leq g(x,s)$ and $R^Z(x,s)$ holds where the use of the latter is $\leq \psi(\langle x,i \rangle)$. Then $\hat{R}$ is ψ-bounded computable and, as one can easily check,

$$\exists s R^{\emptyset'}(x,s) \text{ iff } \exists i < h(x) \exists s \hat{R}(\langle i,x \rangle, s).$$

So, for the $\widehat{\Sigma}_1^{\emptyset'}$-set Y defined by $\hat{R}$, $x \in X$ iff there is a number $i < h(x)$ such that $\langle x,i \rangle \in Y$. So X is truth-table reducible to Y by a disjunctive reduction.

For 3. We build a set A such that $A \in \breve{\Sigma}_1^{\emptyset'}$ and $A \notin \widehat{\Sigma}_1^{\emptyset'}$. In order to guarantee the latter, it suffices to meet the requirements

$$R_e: \; e \in A \text{ iff } \widehat{\Phi}_e^{\emptyset'}(e) \uparrow.$$

(Namely, given any set $X \in \widehat{\Sigma}_1^{\emptyset'}$, by Proposition 2.1 there is an index e such that $X = dom(\widehat{\Phi}_1^{\emptyset'})$. So R_e ensures $A(e) \neq X(e)$.) In order to guarantee the former, we define a Turing functional Ψ, a c.e. set B and a total function g such that $g \leq_{wtt} \emptyset'$ and the use of the functional Ψ is bounded by g, and we let

$$A = dom(\Psi^B). \tag{1}$$

By $B \leq_{wtt} \emptyset'$ this implies that A is the domain of $\widehat{\Psi}^{\emptyset'}$ where $\widehat{\Psi}$ is a Turing functional such that the use of $\widehat{\Psi}$ is bounded by a total function $\hat{g} \leq_{wtt} \emptyset'$. Obviously, this shows that $A \in \breve{\Sigma}_1^{\emptyset'}$.

In the remainder of the proof we give computable enumerations $\{\Psi_s^X(x)\}_{s \geq 0}$ and $\{B_s\}_{s \geq 0}$ of a Turing functional Ψ and a c.e. set B, respectively, and define a function $g \leq_{wtt} \emptyset'$ such that Ψ, B and g have the required properties. Here Ψ on input e, $g(e)$ and $B \cap \omega^{[e]}$ are determined by the strategy to meet R_e. Note that, by (1), in order to meet the requirement R_e it suffices to ensure

$$\Psi^B(e) \downarrow \text{ iff } \widehat{\Phi}_e^{\emptyset'}(e) \uparrow . \tag{2}$$

For this sake it suffices to guarantee that, for $s \geq 0$,

$$\Psi_s^{B_s}(e) \downarrow \text{ iff } \widehat{\Phi}_{e,s}^{\emptyset'_s}(e) \uparrow . \tag{3}$$

In order to achieve the latter, it is crucial to note that, for the least stage s_0 such that $\widehat{\Phi}_{e,s_0}^{\emptyset'_{s_0}}(e) \downarrow$, $\varphi_{e,s_0}(e) \downarrow$ and, if there is such a stage s_0, then

$$|\{s : \widehat{\Phi}_{e,s}^{\emptyset'_s}(e) \downarrow \ \& \ \widehat{\Phi}_{e,s+1}^{\emptyset'_{s+1}}(e) \uparrow\}| \leq \varphi_e(e)$$

(since, for any such stage s, a number $< \varphi_e(e)$ has to be enumerated into $\emptyset'$ at stage $s+1$) whence

$$|\{s : \widehat{\Phi}_{e,s}^{\emptyset'_s}(e) \uparrow \ \& \ \widehat{\Phi}_{e,s+1}^{\emptyset'_{s+1}}(e) \downarrow\}| \leq \varphi_e(e) + 1. \tag{4}$$

So if we let

$$g(e) = \begin{cases} \langle e, 0 \rangle + 1 & \text{if } \varphi_e(e) \uparrow, \\ \langle e, \varphi_e(e) + 2 \rangle + 1 & \text{otherwise;} \end{cases}$$

define B_s by letting $B_0 = \emptyset$, enumerating $\langle e, x \rangle$ into B_{s+1} for the least x such that $\langle e, x \rangle \notin B_s$ if $\widehat{\Phi}_{e,s}^{\emptyset'_s}(e) \uparrow$ and $\widehat{\Phi}_{e,s+1}^{\emptyset'_{s+1}}(e) \downarrow$, and letting $B_{s+1} \cap \omega^{[e]}$ otherwise; and, finally, define Ψ on e according to (3) (where, for s such that $\Psi_s^{B_s}(e) \downarrow$, the use of this computation is defined to be $\langle e, x \rangle + 1$ for the least x such that $\langle e, x \rangle \notin B_s$) then $g \leq_{wtt} \emptyset'$, Ψ is g-bounded and (3) holds for all s. So R_e is met. $\square$

We conclude this section by pointing out another structural difference between the $\widehat{\Sigma}$ classes and the classical Σ classes following from Proposition 3.3. If we define the classes $\breve{\Sigma}_n^X$ and $\breve{\Pi}_n^X$ in the obvious way and if we let $\breve{\Sigma}_n^0 = \breve{\Sigma}_n^{\emptyset}$ $(n \geq 1)$ then, just as in the proof of Proposition 2.2, we may show that $\breve{\Sigma}_2^0 = \breve{\Sigma}_1^{\emptyset'}$. Moreover, it is easy to show that the classes $\widetilde{\Sigma}_n^0$ and $\breve{\Sigma}_n^0$ (based on total use bounds) have in common with the classical classes

Σ^0_n that they are downward closed under disjunctive reductions (in fact, for $\widetilde{\Sigma}^0_n$ this is immediate by the stronger Proposition 3.2). By Proposition 3.3 this implies that $\breve{\Sigma}^0_2$ is the downward closure of the class $\widehat{\Sigma}^0_2 = \widehat{\Sigma}^{\emptyset'}_1$ under disjunctive reductions and that the class $\widehat{\Sigma}^0_2$ (based on partial use bounds) fails to be downward closed under such reductions.

4. The Low Basis Theorem

A classical result on Π^0_1-classes is the superlow basis theorem of Jockusch and Soare[12] which states that a nonempty Π^0_1 class has a member of superlow degree, and hence a member A for which $A^\dagger \equiv_{wtt} \emptyset'$ in particular. One of the corollaries to the proof is that coding can also be used. Hence, as stated in[12], if P is a special Π^0_1 class (i.e., P is nonempty and has no computable members) then, for any set S with $\emptyset' \leq_T S$, P contains a member A with $A' \equiv_T S$. We will see in this section that this generalization fails for the *wtt*-jump. Recall the proof of the Generalized Low Basis Theorem.

Let T be a computable tree with $\mathrm{P} = [T]$. Of course T has $2^{\aleph_0}$ many paths as it has no computable paths. We inductively define computable trees T_n such that $T = T_0 \supseteq T_1 \supseteq \cdots$. In the first step, given T_{2e}, as usual let $U_{2e} = \{\sigma \in T_{2e} : \Phi^\sigma_e(e)[|\sigma|] \uparrow\}$. If U_{2e} is finite let $T_{2e+1} = T_{2e}$ else $T_{2e+1} = U_{2e}$ which also has $2^{\aleph_0}$ many paths. Note that this decision is $\leq_{wtt} \emptyset'$, which is the reason why the Low Basis Theorem actually gives a superlow member.

The second step is to find the lex-least pair (σ_0, σ_1) of incomparable strings in T_{2e+1} with $\sigma_0 <_L \sigma_1$, and let T_{2e+2} be the full subtree of T_{2e+1} above $\sigma_{S(e)}$, and then let $A \in \bigcap_e [T_e]$.

It is this second step which fails to work for the bounded jump $A^\dagger$ since we seem to need the full power of a Turing reduction to $\emptyset'$ to find these two strings. This is a fatal problem:

Theorem 4.1. *There is a set S with $\emptyset' <_{wtt} S$ and a computable tree T with $[T] \neq \emptyset$ and no computable path such that, for all $A \in [T]$, $A^\dagger \not\equiv_{wtt} S$.*

Proof. The argument is finite injury. It suffices to define a Δ^0_2 set D and a computable tree T such that the following hold.

$$|[T]| = 2^{\aleph_0} \text{ (hence, in particular, } [T] \neq \emptyset) \tag{5}$$

$$\forall\, A \in [T]\ (A \text{ is not computable}) \tag{6}$$

$$\forall\, A \in [T]\ (D \not\leq_{wtt} A^\dagger). \tag{7}$$

(Then, for instance, the set $S = \emptyset'' \oplus D$ has the required properties.)

The scheme for defining T is as follows. We give a computable approximation $\iota : 2^{<\omega} \times \omega \to 2^{<\omega}$ (where $\lambda\sigma.\iota(\sigma, s)$ is defined at stage s of the construction) of a function $\iota^* : 2^{<\omega} \to 2^{<\omega}$ which induces an isomorphism from 2^ω onto $[T]$ for the desired tree T.

To be more precise, for any string σ and any stage s we guarantee

$$\forall\, i \leq 1\ (\iota(\sigma, s) \prec \iota(\sigma i, s)) \tag{8}$$

and

$$\iota(\sigma 0, s) \text{ and } \iota(\sigma 1, s) \text{ are incomparable,} \tag{9}$$

and we ensure that

$$\iota^*(\sigma) = \lim_{s \to \omega} \iota(\sigma, s) \text{ exists.} \tag{10}$$

Initially we let $\iota(\sigma, 0) = \sigma$ for any string σ. Then, at stage $s+1$, we guarantee

$$range(\lambda\sigma.\iota(\sigma, s+1)) \subseteq range(\lambda\sigma.\iota(\sigma, s)). \tag{11}$$

So, if we let

$$T_s = \{\tau : \exists \sigma(\tau \preceq \iota(\sigma, s))\}$$

then $T_0 = 2^{<\omega}$ and, by computability of ι and by (8) and (9), T_s, $s \geq 0$, are uniformly computable trees such that $[T_s]$ is isomorphic to 2^ω. Moreover, by (11), $T_{s+1} \subseteq T_s$ and, by (10),

$$T = \lim_{s \to \infty} T_s \tag{12}$$

exists and satisfies (5). Note that the tree T might be not computable. But, by (8) and (9), a node τ has an extension in $[T_s]$ iff $\tau \preceq \iota(\sigma, s)$ for some string σ of length $|\tau|$, and, by (10) and (11), τ has an extension in $[T]$ iff τ has extensions in $[T_s]$ for all $s \geq 0$. So the set of nodes which do not have extensions in $[T]$ is c.e., hence $\overline{[T]}$ is a Σ^0_1-class. It follows that there is a computable tree $\hat{T}$ with $[T] = [\hat{T}]$. So it suffices to ensure that the tree T defined by (12) satisfies (6) and (7). (Note that, as observed above already, T satisfies (5).)

For this sake it suffices to meet the requirements

$$R_{2e} : \forall\, A \in [T]\ (A \neq \varphi_e)$$

and

$$R_{2e+1} : \forall\, A \in [T]\ (\widehat{\Phi}_e^{A^\dagger} \neq D),$$

respectively ($e \geq 0$).

The requirement R_n is allowed to change the approximations to $\iota^*(\sigma)$ only for the nodes σ of length $\geq n$. Since the requirements will be finitary, this guarantees (10). The requirement R_n is split into the subrequirements $R_{n,\sigma}$ where $|\sigma| = n$ and where $R_{n,\sigma}$ guarantees that R_n is met for all sets A such that $\iota^*(\sigma)$ is an initial segment of A. By definition of T this ensures that R_n is met. Moreover, whenever $R_{n,\sigma}$ changes the approximation ι at a stage $s+1$, then it moves $\iota(\sigma, s+1)$ to $\iota(\sigma\tau, \sigma)$ for a proper extension $\sigma\tau$ of σ and simultaneously moves $\iota(\sigma', s+1)$ to the corresponding extension $\iota(\sigma'\tau, s)$ of $\iota(\sigma', s)$ for any proper extension σ' of σ (and leaves $\iota(\sigma'', s)$ unchanged for nodes σ'' not extending σ). Obviously, this ensures that ι has the required properties (8), (9) and (11).

The strategy to meet the noncomputability requirement $R_{2e,\sigma}$ for any node σ with $|\sigma| = 2e$ is as follows. Wait for a stage $s+1$, such that $\varphi_{e,s}(x)$ is defined for the least x such that $\iota(\sigma 0, s)(x) \neq \iota(\sigma 1, s)(x)$ (note that, by (9) and (8), such an x exists and $x \geq |\iota(\sigma, s)|$; also note that if there is no such stage s then, by (10), φ_e is not total and R_{2e} is trivially met). Then fix $i \leq 1$ minimal such that $\iota(\sigma i, s)(x) \neq \varphi_e(x)$ and let $\iota(\sigma\tau, s+1) = \iota(\sigma i\tau, s)$ for all strings τ (and let $\iota(\tilde{\sigma}, s{+}1) = \iota(\tilde{\sigma}, s)$ for any node $\tilde{\sigma}$ such that $\sigma \not\preceq \tilde{\sigma}$). If no higher priority requirement R_n ($n < 2e$) acts after stage s this will guarantee that $\iota^*(\sigma) = \iota(\sigma i, s)$ hence $A(x) \neq \varphi_e(x)$ for all sets A extending $\iota^*(\sigma)$.

The strategy to meet the jump requirement $R_{2e+1,\sigma}$ for any node σ with $|\sigma| = 2e+1$ is as follows. First we fix a number x_σ to be reserved for this subrequirement. Next we will define $\iota^*(\sigma)$ in such a way that if there is a set A such that $\iota^*(\sigma) \prec A$ and $\widehat{\Phi}_e^{A^\dagger}(x_\sigma)$ is defined then, for all such sets A, the value of $\widehat{\Phi}_e^{A^\dagger}(x_\sigma)$ depends only on $\iota^*(\sigma)$. So, finally, we will meet R_{2e+1} for A extending $\iota^*(\sigma)$ by letting $D(x_\sigma)$ diagonalize against this value. The details are as follows.

Wait for a stage s_σ such that $\varphi_{e,s_\sigma}(x_\sigma)\downarrow$ (if there is no such stage then $\widehat{\Phi}_e^{A^\dagger}(x_\sigma)\uparrow$ for all sets A hence R_{2e+1} is trivially met). Then, for any stage $s \geq s_\sigma$, let

$$U(\sigma, s) = \{z : z < \varphi_e(x_\sigma) \ \&\ \varphi_{z,s}(z)\downarrow\} \text{ and } u(\sigma, s) = \max\{\varphi_z(z) : z \in U(\sigma, s)\}.$$

If $|\iota(\sigma, s)| < u(\sigma, s)$ then fix the least (in the sense of the length-lexicographic order) σ' extending σ such that $|\iota(\sigma', s)| \geq u(\sigma, s)$ and let $\iota(\sigma\tau, s+1) = \iota(\sigma'\tau, s)$ for all strings τ (and let $\iota(\tilde{\sigma}, s+1) = \iota(\tilde{\sigma}, s)$ for any node $\tilde{\sigma}$ such that $\sigma \not\preceq \tilde{\sigma}$). Note that there are at most $\varphi_e(x_\sigma)$ stages s at which $u(\sigma, s)$ may grow. So this action is finitary (hence compatible with

(10)). Finally, define the computable approximation $D(x_\sigma, s+1)$ of $D(x_\sigma)$ at stage $s+1 > s_\sigma$ as follows (and let $D(x_\sigma, s) = 0$ if $s \leq s_\sigma$ or s_σ is not defined). For any set A let $A^\dagger_s = \{z : \varphi_{z,s}(z) \downarrow \ \& \ \widehat{\Phi}^A_{z,s}(z) \downarrow\}$. Then $D(x_\sigma, s+1) = 1$ if $\widehat{\Phi}^{(\iota(\sigma,s+1))^\dagger}_{e,s}(x_\sigma) \downarrow = 0$ and $D(x_\sigma, s+1) = 0$ otherwise (where we view $\iota(\sigma, s+1)$ as the set of the numbers $y < |\iota(\sigma, s+1)|$ such that $\iota(\sigma, s+1)(y) = 1$). Note that, by (10), $\widehat{\Phi}^{(\iota(\sigma,s+1))^\dagger}_{e,s}(x_\sigma) = \widehat{\Phi}^{(\iota^*(\sigma))^\dagger}_{e}(x_\sigma)$ for all sufficiently large s, hence $D(x_\sigma) = \lim_{s\to\omega} D(x_\sigma, s)$ exists.

In order to show that the above strategy succeeds to meet $R_{2e+1,\sigma}$, w.l.o.g. assume that $\varphi_e(x_\sigma) \downarrow$ and that A is a set extending $\iota^*(\sigma)$ such that $\widehat{\Phi}^{A^\dagger}_e(x_\sigma) \downarrow$. Since $D(x_\sigma) = \lim_{s\to\omega} D(x_\sigma, s)$, it suffices to show that $D(x_\sigma, s+1) \neq \widehat{\Phi}^{A^\dagger}_e(x_\sigma)$ for all sufficiently large s. So fix s large enough such that $s \geq s_\sigma$, $\iota(\sigma, s+1) = \iota^*(\sigma)$, $U(\sigma, s) = \{z < \varphi_e(x_\sigma) : \varphi_z(z) \downarrow\}$ and $\widehat{\Phi}^{A^\dagger}_{e,s}(x_\sigma) \downarrow$. Since, by $s \geq s_\sigma$ and by construction, $D(x_\sigma, s+1) \neq \widehat{\Phi}^{(\iota(\sigma,s+1))^\dagger}_{e,s}(x_\sigma)$, it suffices to show that $(\iota(\sigma, s+1))^\dagger \restriction \varphi_e(x_\sigma) = A^\dagger \restriction \varphi_e(x_\sigma)$. So fix $z < \varphi_e(x_\sigma)$. If $\varphi_z(z) \uparrow$ then $(\iota(\sigma, s+1))^\dagger(z) = A^\dagger(z) = 0$. So we may assume $\varphi_z(z) \downarrow$. Then, by choice of s and by construction, $\varphi_z(z) \leq u(\sigma, s) \leq |\iota(\sigma, s+1)|$. Since $\iota(\sigma, s+1) = \iota^*(\sigma) \prec A$, it follows that $\iota(\sigma, s+1) \restriction \varphi_z(z) = A \restriction \varphi_z(z)$. So $\widehat{\Phi}^{\iota(\sigma,s+1)}_z(z) \downarrow$ iff $\widehat{\Phi}^A_z(z) \downarrow$, i.e., $(\iota(\sigma, s+1))^\dagger(z) = A^\dagger(z)$.

This completes the description of the individual strategies. The rest is a straightforward application of the finite injury method. □

References

1. K. Ambos-Spies, R. Downey and M. Monath, On the computational power of maximal sets, in preparation.
2. B. Anderson and B. Csima, A bounded jump for the bounded Turing degrees, *Notre Dame J. Form. Log.* 55 (2014), no. 2, 245–264.
3. B. Afshari, G. Barmpalias, S. B. Cooper and F. Stephan, Post's programme for the Ershov hierarchy, *J. Logic Comput.* 17 (2007), no. 6, 1025–1040.
4. G. Barmpalias, R. Downey and N. Greenberg, Working with strong reducibilities above totally ω-c.e. and array computable degrees, *Trans. Amer. Math. Soc.* 362 (2010), no. 2, 777–813.
5. J. Chisholm, J. Chubb, V. Harizanov, D. Hirschfeldt, C. Jockusch, Jr., T. McNicholl, and S. Pingrey, Π^0_1 classes and strong degree spectra of relations *J. Symbolic Logic* 72 (2007), no. 3, 1003–1018.
6. R. Coles, R. Downey and G. LaForte, Notes on wtt-jump and ordinal notations. Manuscript, 1998.
7. B. Csima, R. Downey and K. M. Ng, Limits on jump inversion for strong reducibilities, *J. Symbolic Logic* 76 (2011), no. 4, 1287–1296.
8. R. Downey and N. Greenberg, A Hierarchy of Turing Degrees, in Series:

Annals of Mathematics Studies, Vol. 206, Princeton University Press 2020, 240 pages, ISBN 9780691199658.

9. R. Downey and D. Hirschfeldt, *Algorithmic Randomness and Complexity*, Springer-Verlag, 2010.
10. Y. Ershov, A certain hierarchy of sets. I. (Russian), *Algebra i Logika* 7 (1968), no. 1, 47–74.
11. G. Gerla, Una generalizzazione della gerarchia di Ershov, *Bolletino U.M.I.* (5) 16-B (1979), no. 2 765–778.
12. C. Jockusch and R. Soare, Degrees of members of Π_1^0 classes, *Pacific J. Math.* 40 (1972), 605–616.
13. S. Kleene and E. Post, The upper semi-lattice of degrees of recursive unsolvability, *Ann. of Math. (2)* 59 (1954), 379–407.
14. J. Mohrherr, Density of a final segment of the truth-table degrees, *Pacific J. Math.* 115 (1984), no. 2, 409–419.
15. P. Odifreddi, Classical Recursion Theory, Vol. II, North-Holland, 1999.
16. G. Sacks, On the degrees less than $\mathbf{0}'$, *Ann. of Math. (2)* 77 (1963), no. 2, 211–231.
17. A. Turing, Systems of logic based on ordinals, *Proc. London Math. Soc. (2)* 45 (1939), no. 3, 161–228.
18. G. Wu and H. Wu, Bounded jump and the high/low hierarchy, in *Theory and applications of models of computation, Lecture Notes in Comput. Sci., 11436, Springer, Cham, 2019*, 647–658.

E-combinations and closures of quite o-minimal theories

B. Sh. Kulpeshov
Institute of Mathematics and Mathematical Modeling, Shevchenko street 28, Almaty, 050010, Kazakhstan
Kazakh-British Technical University, ul. Tole bi, 59, Almaty, 050000, Kazakhstan
Novosibirsk State Technical University, pr. K. Marx, 20, Novosibirsk, 630073, Russia
E-mail: kulpesh@mail.ru

S. V. Sudoplatov
Sobolev Institute of Mathematics SB RAS, pr. Acad. Koptyuga, 4, Novosibirsk, 630090, Russia
Novosibirsk State Technical University, pr. K. Marx, 20, Novosibirsk, 630073, Russia
Novosibirsk State University, ul. Pirogova, 1, Novosibirsk, 630090, Russia
E-mail: sudoplat@math.nsc.ru

We consider E-combinations of quite o-minimal theories and characterize preservations of $\aleph_0$-categoricity for these combinations. We describe closures and generating sets for families of quite o-minimal theories with respect to the E-operator. Values for e-spectra of these families are found.

Keywords: Quite o-minimal theory, E-combination, $\aleph_0$-categoricity, Ehrenfeuchtness, family of theories, closure, e-spectrum.

1. Preliminaries

In a series of papers[1–7] topological properties for families of theories are studied. The notions of P-operator and E-operator were introduced allowing to study links between theories with respect to appropriate closure operators. These operators give possibilities to generate new theories by given families of ones, and, in special cases, to find minimal/least generating sets. Counting e-spectra for families of theories we get characteristics for maximal variations of theories in closed sets of theories with respect to generating sets.

We continue this investigation applying for families of ordered, in particular, quite o-minimal theories[8]. We consider combinations of ordered theories and characterize preservations of countable categoricity for combinations. We describe closures and generating sets for families of quite o-minimal theories with respect to the E-operator.

Let E be an equivalence relation, $(\mathcal{A}_i)_{i\in I}$ be a family of disjoint structures of relational languages such that every E-class is the universe of some $\mathcal{A}_i$, $i \in I$. The structure $\bigcup\limits_{i\in I} \mathcal{A}_i$, if it exists, expanded by the binary predicate $E \notin \Sigma = \bigcup\limits_{i\in I} \Sigma(\mathcal{A}_i)$ is the *E-union* of the structures $\mathcal{A}_i$. Here we assume that if $R \in \Sigma \setminus \Sigma(\mathcal{A}_i)$ then R is interpreted on the universe A_i of $\mathcal{A}_i$ to be empty, and common symbols $R \in \Sigma(\mathcal{A}_i) \cap \Sigma(\mathcal{A}_j)$ have same arities. The resulted structure of the language $\Sigma \cup \{E\}$ is denoted $\mathcal{A}_E$ and the operator mapping $(\mathcal{A}_i)_{i\in I}$ to $\mathcal{A}_E$ is the *E-operator*.

Notice that the definition for the structure $\mathcal{A}_E$ can be naturally spread for $\mathcal{A}_i$ with functional language symbols f replacing interpretations for f by their graphs Γ_f, in particular, replacing constants by singletons.

The structure $\mathcal{A}_E$ is called the *E-combination* of the structures $\mathcal{A}_i$ and denoted by $\mathrm{Comb}_E(\mathcal{A}_i)_{i\in I}$ if $\mathcal{A}_i = (\mathcal{A}_E|_{A_i})|_{\Sigma(\mathcal{A}_i)}$, $i \in I$. Structures $\mathcal{A}'$, which are elementarily equivalent to $\mathrm{Comb}_E(\mathcal{A}_i)_{i\in I}$, will be also considered as E-combinations.

Clearly, there are E-combinations such that all $\mathcal{A}' \equiv \mathcal{A}_E$ can be represented as E-combinations of some $\mathcal{A}'_j \equiv \mathcal{A}_i$. We call this representability of $\mathcal{A}'$ to be the *E-representability.*

If there is $\mathcal{A}' \equiv \mathcal{A}_E$ which is not E-representable, we have the E'-representability replacing E by E' such that E' is obtained from E adding equivalence classes with models for all theories T, where T is a theory of a restriction $\mathcal{B}$ of a structure $\mathcal{A}' \equiv \mathcal{A}_E$ to some E-class and $\mathcal{B}$ is not elementarily equivalent to the structures $\mathcal{A}_i$. The resulting structure $\mathcal{A}_{E'}$ (with the E'-representability) is a *e-completion*, or a *e-saturation*, of $\mathcal{A}_E$. The structure $\mathcal{A}_{E'}$ itself is called *e-complete*, or *e-saturated*, or *e-universal*, or *e-largest.*

For a structure $\mathcal{A}_E$ the number of *new* structures with respect to the structures $\mathcal{A}_i$, i. e., of the structures $\mathcal{B}$, forming E-classes of $\mathcal{A}' \equiv \mathcal{A}_E$, which are pairwise elementarily non-equivalent and elementarily non-equivalent to the structures $\mathcal{A}_i$, is called the *e-spectrum* of $\mathcal{A}_E$ and denoted by e-$\mathrm{Sp}(\mathcal{A}_E)$. The value $\sup\{e\text{-}\mathrm{Sp}(\mathcal{A}') \mid \mathcal{A}' \equiv \mathcal{A}_E\}$ is called the *e-spectrum* of the theory $\mathrm{Th}(\mathcal{A}_E)$ and denoted by e-$\mathrm{Sp}(\mathrm{Th}(\mathcal{A}_E))$.

If $\mathcal{A}_E$ does not have E-classes $\mathcal{A}_i$, which can be removed, with all E-classes $\mathcal{A}_j \equiv \mathcal{A}_i$, preserving the theory $\mathrm{Th}(\mathcal{A}_E)$, then $\mathcal{A}_E$ is called *e-prime*, or *e-minimal.*

For a structure $\mathcal{A}' \equiv \mathcal{A}_E$ we denote by $\mathrm{TH}(\mathcal{A}')$ the set of all theories $\mathrm{Th}(\mathcal{A}_i)$ of E-classes $\mathcal{A}_i$ in $\mathcal{A}'$.

By the definition, an e-minimal structure $\mathcal{A}'$ consists of E-classes with a minimal set $\mathrm{TH}(\mathcal{A}')$. If $\mathrm{TH}(\mathcal{A}')$ is the least for models of $\mathrm{Th}(\mathcal{A}')$ then $\mathcal{A}'$ is called *e-least*.

Definition 1.1.[2] Let $\overline{\mathcal{T}}$ be the class of all complete elementary theories of relational languages. For a set $\mathcal{T} \subset \overline{\mathcal{T}}$ we denote by $\mathrm{Cl}_E(\mathcal{T})$ the set of all theories $\mathrm{Th}(\mathcal{A})$, where $\mathcal{A}$ is a structure of some E-class in $\mathcal{A}' \equiv \mathcal{A}_E$, $\mathcal{A}_E = \mathrm{Comb}_E(\mathcal{A}_i)_{i\in I}$, $\mathrm{Th}(\mathcal{A}_i) \in \mathcal{T}$. As usual, if $\mathcal{T} = \mathrm{Cl}_E(\mathcal{T})$ then $\mathcal{T}$ is said to be *E-closed*.

The operator Cl_E of E-closure can be naturally extended to the classes $\mathcal{T} \subset \overline{\mathcal{T}}$ as follows: $\mathrm{Cl}_E(\mathcal{T})$ is the union of all $\mathrm{Cl}_E(\mathcal{T}_0)$ for subsets $\mathcal{T}_0 \subseteq \mathcal{T}$.

For a set $\mathcal{T} \subset \overline{\mathcal{T}}$ of theories in a language Σ and for a sentence φ with $\Sigma(\varphi) \subseteq \Sigma$ we denote by $\mathcal{T}_\varphi$ the set $\{T \in \mathcal{T} \mid \varphi \in T\}$.

Theorem 1.1. [2] *If $\mathcal{T} \subset \overline{\mathcal{T}}$ is an infinite set and $T \in \overline{\mathcal{T}} \setminus \mathcal{T}$ then $T \in \mathrm{Cl}_E(\mathcal{T})$ (i.e., T is an accumulation point for $\mathcal{T}$ with respect to E-closure Cl_E) if and only if for any formula $\varphi \in T$ the set $\mathcal{T}_\varphi$ is infinite.*

Theorem 1.2. [2] *If $\mathcal{T}_0'$ is a generating set for a E-closed set $\mathcal{T}_0$ then the following conditions are equivalent:*

(1) *$\mathcal{T}_0'$ is the least generating set for $\mathcal{T}_0$;*

(2) *$\mathcal{T}_0'$ is a minimal generating set for $\mathcal{T}_0$;*

(3) *any theory in $\mathcal{T}_0'$ is isolated by some set $(\mathcal{T}_0')_\varphi$, i.e., for any $T \in \mathcal{T}_0'$ there is $\varphi \in T$ such that $(\mathcal{T}_0')_\varphi = \{T\}$;*

(4) *any theory in $\mathcal{T}_0'$ is isolated by some set $(\mathcal{T}_0)_\varphi$, i.e., for any $T \in \mathcal{T}_0'$ there is $\varphi \in T$ such that $(\mathcal{T}_0)_\varphi = \{T\}$.*

Recall that a theory T is *Ehrenfeucht* if T has finitely many countable models ($I(T,\omega) < \omega$) but is not ω-categorical ($I(T,\omega) > 1$). A structure with an Ehrenfeucht theory is also *Ehrenfeucht*.

Now we consider the following criteria for countable categoricity and Ehrenfeuchtness of disjoint E-combinations. These criteria are formulated for the case when combinations are not ordered. Below, in the following sections, we will consider similar criteria for disjoint combinations in the ordered case.

Theorem 1.3. [1] *If the languages $\Sigma(\mathcal{A}_i)$ are at most countable, $i \in I$, $|I| \leq \omega$, and the structure $\mathcal{A}_E$ is infinite then the theory $\mathrm{Th}(\mathcal{A}_E)$ is ω-categorical if and only if each structure $\mathcal{A}_i$ is either finite or ω-categorical, and I is either finite, or infinite and E_i-classes do not approximate infinitely many n-types for any $n \in \omega$.*

Theorem 1.4. [1] *If the language* $\bigcup_{i\in I} \Sigma(\mathcal{A}_i)$ *is at most countable and the structure* $\mathcal{A}_E$ *is infinite then the theory* $T = \mathrm{Th}(\mathcal{A}_E)$ *is Ehrenfeucht if and only if* e-Sp$(T) < \omega$ *(which is equivalent here to* e-Sp$(T) = 0$*) and for an e-largest model* $\mathcal{A}_{E'} \models T$ *consisting of* E'*-classes* $\mathcal{A}_j$, $j \in J$, *the following conditions hold:*

(a) *for any* $j \in J$, $I(\mathrm{Th}(\mathcal{A}_j), \omega) < \omega$;

(b) *there are positively and finitely many* $j \in J$ *such that* $I(\mathrm{Th}(\mathcal{A}_j), \omega) > 1$;

(c) *if* $I(\mathrm{Th}(\mathcal{A}_j), \omega) \leq 1$ *then there are always finitely many* $\mathcal{A}_{j'} \equiv \mathcal{A}_j$ *or always infinitely many* $\mathcal{A}_{j'} \equiv \mathcal{A}_j$ *independent of* $\mathcal{A}_{E'} \models T$.

Recall that a subset A of a linearly ordered structure M is called *convex* if for any $a, b \in A$ and $c \in M$ whenever $a < c < b$ we have $c \in A$. A *weakly o-minimal structure* is a linearly ordered structure $M = \langle M, =, <, \ldots\rangle$ such that any definable (with parameters) subset of the structure M is a union of finitely many convex sets in M.

In the following definitions M is a linearly ordered structure, $A, B \subseteq M$, M be $|A|^+$-saturated, $p, q \in S_1(A)$ be non-algebraic types.

Definition 1.2. We say that p is *weakly orthogonal* to q ($p \perp^w q$) if $p(x) \cup q(y)$ has an unique extension to a complete 2-type over A.

Definition 1.3.[8] We say that p is *quite orthogonal* to q ($p \perp^q q$) if there is no an A-definable bijection $f : p(M) \to q(M)$. We say that a weakly o-minimal theory is *quite o-minimal* if the notions of weak and quite orthogonality of 1-types coincide.

Definition 1.4. [9] Let T be a weakly o-minimal theory, $\mathcal{M}$ be a sufficiently saturated model of T, and let $\phi(x)$ be an arbitrary M-definable formula with one free variable. The *convexity rank of the formula* $\phi(x)$ ($RC(\phi(x))$) is defined as follows:

1) $RC(\phi(x)) \geq 1$ if $\phi(\mathcal{M})$ is infinite.

2) $RC(\phi(x)) \geq \alpha + 1$ if there are a parametrically definable equivalence relation $E(x, y)$ and infinitely many elements $b_i, i \in \omega$, such that:

- For any $i, j \in \omega$ whenever $i \neq j$ we have $\mathcal{M} \models \neg E(b_i, b_j)$;
- For each $i \in \omega$ $\quad RC(E(x, b_i)) \geq \alpha$ and $E(\mathcal{M}, b_i)$ is a convex subset of $\phi(\mathcal{M})$.

3) $RC(\phi(x)) \geq \delta$ if $RC(\phi(x)) \geq \alpha$ for all $\alpha < \delta$ (δ is limit).

If $RC(\phi(x)) = \alpha$ for some α then we say that $RC(\phi(x))$ is defined. Otherwise (i.e. if $RC(\phi(x)) \geq \alpha$ for all α), we put $RC(\phi(x)) = \infty$.

The *convexity rank of an* 1-*type* p $(RC(p))$ is called the infimum of the set $\{RC(\phi(x)) \mid \phi(x) \in p\}$, i.e. $RC(p) := \inf\{RC(\phi(x)) \mid \phi(x) \in p\}$.

We say that a quite o-minimal theory T has *few countable models* if T has less than 2^ω pairwise non-isomorphic countable models. In[10] quite o-minimal theories with few countable models were studied. In particular, the countable spectrum of quite o-minimal theories has been described. This result generalizes the theorem on the countable spectrum for o-minimal theories obtained by L. Mayer in[11].

Theorem 1.5. [10] *Let T be a quite o-minimal theory in a countable language. Then either T has 2^ω countable models or T has exactly $6^n 3^m$ countable models, where n and m are non-negative integers. Moreover, for any non-negative integers n, m there is a quite o-minimal theory T with exactly $6^n 3^m$ countable models.*

In Section 2 we present a criterion for quite o-minimality of an E-combination of countably many copies of an $\aleph_0$-categorical quite o-minimal structure (Theorem 2.1). We also obtain that the elementary theory of such an E-combination in case of quite o-minimality is either $\aleph_0$-categorical or has 2^ω countable models (Corollary 2.1).

In Section 3, e-spectra for families of quite o-minimal theories are described (Theorem 3.1, Proposition 3.4).

2. E-combinations

Throughout this section we consider a linearly ordered structure $M := \langle M; <_M, \Sigma\rangle$, where Σ does not contain any constant names.

We denote by $dcl^<_M(\emptyset)$ the set of elements of M which are $\emptyset$-definable by the order relation $<_M$. We also denote by M^+ a linearly ordered disjoint E-combination of ω copies of M in the language $\{<, \Sigma, E^2, c_i^k\}_{i<\lambda, k\in\omega}$, i.e. the universe of the combination is $\bigcup_{k\in\omega} M_k$; M_k is a copy of M for each $k \in \omega$ and $\{c_i^k \mid i < \lambda\} \subseteq dcl^<_{M_k}(\emptyset)$ for some ordinal λ; either $M_k < M_m$ or $M_m < M_k$ for any $k, m \in \omega$ with $k \neq m$, and E is an equivalence relation partitioning M^+ into convex classes so that for any $a \in M^+$, $E(a, M^+) = M_k$ for some $k \in \omega$.

Thus, we include in the signature of an E-combination M^+ all the elements lying in $dcl^<_{M_k}(\emptyset)$ for each $k \in \omega$.

Lemma 2.1. *Let M be an $\aleph_0$-categorical quite o-minimal structure, M^+ be a linearly ordered disjoint E-combination of ω copies of M. Then*

$$M^+ \models \forall x \exists y_1 \exists y_2 [y_1 \neq x \wedge y_2 \neq x \wedge y_1 < y_2 \wedge E(x, y_1) \wedge$$

$$E(x, y_2) \wedge \forall t(y_1 \leq t \leq y_2 \to \neg\theta(t))],$$

where $\theta(M^+)$ is the set of elements of M^+ having an immediate predecessor or an immediate successor.

Proof of Lemma 2.1. Let $M_<$ be the reduct of M on $\{<\}$. By $\aleph_0$-categoricity of M the structure $M_< \equiv F_1^{k_1} + \mathbb{Q} + F_2^{k_2} + \mathbb{Q} + \cdots + \mathbb{Q} + F_n^{k_n}$ for some $2 \leq n < \omega$, where $F_i^{k_i}$ is a finite linear ordering consisting of k_i elements, $k_1 \geq 0$, $k_n \geq 0$, and for each $2 \leq i \leq n-1$ if $k_i \neq 0$ then $k_i \geq 2$. Therefore, every E-class contains at least one dense part.

Example 2.1. Let $M = \langle M, < \rangle$ be a linearly ordered structure, where $M = 1 + \mathbb{Q}$. Obviously, M is an $\aleph_0$-categorical o-minimal structure, and $dcl_M^<(\emptyset) = \{a\}$, where a is the first element of M. Let M^+ be a linearly ordered disjoint E-combination of ω copies of M ordered by type ω. Since $dcl_M^<(\emptyset) \neq \emptyset$, every E-class in M^+ has an own name, i.e. it equals $E(c_i, M^+)$ for some $c_i \in dcl_{M^+}(\emptyset)$. Consider the following set of formulas:

$$C(x) := \{\forall y[E(c_i, y) \to y < x] \mid i \in \omega\}.$$

Obviously, $C(x)$ determines an $\emptyset$-definable cut in M^+.

We assert that $Th(M^+)$ has 2^ω countable models. Indeed, let $\mathbb{ZE}$ be the set of E-classes ordered by type $\omega^* + \omega$. Also, recall that F^k denotes a finite linear ordering consisting of k elements. Then we denote by $F^{k,\mathbb{ZE}}$ ($\omega^{\mathbb{ZE}}$ and $\mathbb{Q}^{\mathbb{ZE}}$) the set of $\mathbb{ZE}$-copies ordered by type F^k (ω and $\mathbb{Q}$ respectively). We see that $C(x)$ can be realized by $\mathbb{Q}^{\mathbb{ZE}}$, $F^{k,\mathbb{ZE}}$, $\omega^{\mathbb{ZE}}$, $F^{k,\mathbb{ZE}} + \mathbb{Q}^{\mathbb{ZE}}$, $\mathbb{Q}^{\mathbb{ZE}} + F^{k,\mathbb{ZE}}$ or $F_1^{k_1,\mathbb{ZE}} + \mathbb{Q}^{\mathbb{ZE}} + F_2^{k_2,\mathbb{ZE}} + \mathbb{Q}^{\mathbb{ZE}} + \cdots + F_n^{k_n,\mathbb{ZE}} + \mathbb{Q}^{\mathbb{ZE}}$ for any $k < \omega$, $0 \leq n, k_i \leq \omega$, where for each $2 \leq i \leq n-1$ if $k_i \neq 0$ then $k_i \geq 2$; and if $k_i = \omega$ then $F_i^{k_i,\mathbb{ZE}} \equiv \omega^{\mathbb{ZE}}$.

Proposition 2.1. *Let M be an $\aleph_0$-categorical quite o-minimal structure, M^+ be a linearly ordered disjoint E-combination of ω copies of M. Suppose that $dcl_M^<(\emptyset) \neq \emptyset$. Then $Th(M^+)$ has 2^ω countable models.*

Proof of Proposition 2.1. Since $dcl_M^<(\emptyset) \neq \emptyset$, by $\aleph_0$-categoricity of M there are finitely many $c \in dcl_M^<(\emptyset)$. Consequently, every E-class in M^+ has an own name, i.e. it equals $E(c_i, M^+)$ for some $c_i \in dcl_{M^+}(\emptyset)$.

Case 1. There exists at least one infinite discretely ordered chain $\{E(c_i, x) \mid i < \omega\}$ of E-classes in M^+.

Without loss of generality, suppose that this chain is increasing. Then we have two possibilities:

(1) For any $a \in M^+$ there is $i < \omega$ such that $a < E(c_i, M^+)$;

(2) There is $c' \in dcl_{M^+}(\emptyset)$ such that $E(c_i, M^+) < E(c', M^+)$ for each $i < \omega$.

Without loss of generality, suppose the first. Then

$$C(x) := \{\forall y[E(c_i, y) \to y < x] \mid i \in \omega\}$$

determines an $\emptyset$-definable cut in M^+.

Then using the notations in Example 2.1 we see that $C(x)$ can be realized by

$$F_1^{k_1,\mathbb{ZE}} + \mathbb{Q}^{\mathbb{ZE}} + F_2^{k_2,\mathbb{ZE}} + \mathbb{Q}^{\mathbb{ZE}} + \cdots + F_n^{k_n,\mathbb{ZE}} + \mathbb{Q}^{\mathbb{ZE}}$$

for any $0 \le n, k_i \le \omega$, where for each $2 \le i \le n-1$ if $k_i \ne 0$ then $k_i \ge 2$; and if $k_i = \omega$ then $F_i^{k_i,\mathbb{ZE}} \equiv \omega^{\mathbb{ZE}}$, whence $Th(M^+)$ has 2^ω countable models.

Case 2. There is no infinite discretely ordered chain of E-classes in M^+.

Then there is $n < \omega$ such that M^+ contains n-orderings of E-classes and there is no m-orderings of E-classes for any $m > n$. Consequently, there exist $n_1 < \omega$ with $1 \le n_1 \le n$ and an interval (a, b) in M^+ on which these n_1-orderings are densely ordered. Then we take an arbitrary E-class $E(c_i, M^+)$ containing in (a, b) and being the last element of the n_1-ordering and consider the following set of formulas:

$$C(x) := \{\neg E(c_i, x) \wedge c_i < x\} \cup \{\forall y[E(c_j, y) \to x < y] \mid c_i < c_j \wedge \neg E(c_i, c_j)\}.$$

Obviously, $C(x)$ determines an $\emptyset$-definable cut in M^+. By dense ordering of n_1-orderings in (a, b) we have 2^ω $\emptyset$-definable cuts, whence $Th(M^+)$ is not small, and consequently $Th(M^+)$ has 2^ω countable models.

Theorem 2.1. *Let M be an $\aleph_0$-categorical quite o-minimal structure non-having distinguished constants, M^+ be a linearly ordered disjoint E-combination of ω copies of M. Then $Th(M^+)$ is quite o-minimal iff the following holds:*

(1) M^+/E is partitioned into finitely many convex sets on each of which either all the elements have both an immediate predecessor and an immediate successor or all the elements don't have an immediate predecessor or an immediate successor.

(2) M is 1-indiscernible.

Proof of Theorem 2.1. ($\Rightarrow$) Consider the following formula:

$$\theta(x) := \exists t \exists y[t < x < y \wedge \neg E(t,x) \wedge \neg E(x,y) \wedge \forall u \forall z(t < u < x < z < y \rightarrow$$

$$\rightarrow (E(t,u) \vee E(y,x)) \wedge (E(x,z) \vee E(z,y))].$$

Since $Th(M^+)$ is quite o-minimal, $\theta(M^+)$ is a union of finitely many convex sets, and therefore (1) holds.

Prove now that M is dense. If M is not dense then there are elements of M having an immediate predecessor or an immediate successor. By $\aleph_0$-categoricity of M there is $n < \omega$ such that the length of any discretely ordered chain having at least two elements is less than n. By weak o-minimality of M there exist only finitely many such chains, whence there exist only finitely many elements in M having an immediate predecessor or an immediate successor.

Consider the following formula:

$$\phi(x) := \exists y[x < y \wedge \forall z(x \leq z \leq y \rightarrow x = z \vee z = y)].$$

Obviously, $\phi(M^+)$ is a union of infinitely many $\neg\phi(M^+)$-separable convex sets, whence M^+ is not weakly o-minimal, contradicting the hypotheses of the proposition.

Prove now that M has no endpoints. Indeed, if we suppose that there is a left endpoint of M then consider the following formula: $\psi(x) := \forall y[E(x,y) \rightarrow x \leq y]$. Then $\psi(M^+)$ is a union of infinitely many $\neg\psi(M^+)$-separable convex sets, contradicting weak o-minimality of M^+. Similarly there are no right endpoints.

Assume now that M is not 1-indiscernible. Since M has no distinguished constants, there is a convex formula $U(x)$ such that $U(M) \subset M$ and $U(M)$ has no endpoints. But then $U(M^+)$ is a union of infinitely many $\neg U(M^+)$-separable convex sets, contradicting weak o-minimality of M^+.

($\Leftarrow$) Let M be an $\aleph_0$-categorical 1-indiscernible quite o-minimal structure. Then there is $n < \omega$ such that $Th(M)$ has convexity rank n, i.e. there are $\emptyset$-definable equivalence relations $E_1(x,y), \ldots, E_{n-1}(x,y)$ partitioning M into infinitely many infinite convex classes so that $E_1(a,M) \subset E_2(a,M) \subset \cdots \subset E_{n-1}(a,M)$ for any $a \in M$. By Theorem 3 of[12] $Th(M)$ admits quantifier elimination to the language

$$\{=,<\} \cup \{E_i(x,y) \mid 1 \leq i \leq n-1\}.$$

Let F be the set of all finite linear orderings, and $G = F \cup \{\omega, \omega^*, \omega + \omega^*, \omega^* + \omega, \mathbb{Q}\}$. By (1) $M^+ = C_1 + \cdots + C_m$, where C_i/E is elementarily equivalent to some member of G for each $i \leq m$ and m is minimal. Obviously, each C_i is $\emptyset$-definable and consequently there are $\emptyset$-definable formulas $C_i(x), i \leq m$ such that $C_i = C_i(M^+)$.

Consider the following formulas:

[1] Let $F_i(x)$ be the formula which defines the ith finite ordering in M^+/E, $1 \leq i \leq k_1$, where k_1 is the number of all finite orderings in this sum. Denote by n_i the cardinality of the ith finite ordering. Then

$$F_i(x) := \exists x_1 \cdots \exists x_{n_i}[\wedge_{1\leq l,j\leq n_i,l\neq j} \neg E(x_l, x_j) \wedge \wedge_{1\leq l<j\leq n_i} x_l < x_j \wedge$$

$$\wedge \forall t(t < x_1 \wedge \neg E(t, x_1) \rightarrow \exists u[t < u < x_1 \wedge \neg E(t, u) \wedge \neg E(u, x_1)]) \wedge$$

$$\wedge \forall z(x_{n_i} < z \wedge \neg E(x_{n_i}, z) \rightarrow \exists v[x_{n_i} < v < z \wedge \neg E(x_{n_i}, v) \wedge \neg E(v, z)]) \wedge$$

$$\wedge \wedge_{j=1}^{n_i-1} \forall y(x_j < y < x_{j+1} \rightarrow E(x_j, y) \vee E(y, x_{j+1})) \wedge$$

$$(x_1 < x < x_{n_i} \vee [x < x_1 \rightarrow E(x, x_1)] \vee [x_{n_i} < x \rightarrow E(x_{n_i}, x)])].$$

[2] Let $\omega_i(x)$ be the formula which defines the ith ordering ω in M^+/E, $1 \leq i \leq k_2$, where k_2 is the number of all orderings of type ω.

[3] Let $\omega_i^*(x)$ be the formula which defines the ith ordering ω^* in M^+/E, $1 \leq i \leq k_3$, where k_3 is the number of all orderings of type ω^*.

[4] Let $(\omega+\omega^*)_i(x)$ be the formula which defines the ith ordering $\omega+\omega^*$ in M^+/E, $1 \leq i \leq k_4$, where k_4 is the number of all orderings of type $\omega + \omega^*$.

[5] Let $(\omega^*+\omega)_i(x)$ be the formula which defines the ith ordering $\omega^*+\omega$ in M^+/E, $1 \leq i \leq k_5$, where k_5 is the number of all orderings of type $\omega^* + \omega$.

[6] Let $\mathbb{Q}_i(x)$ be the formula which defines the ith ordering of type $\mathbb{Q}$ in M^+/E, $1 \leq i \leq k_6$, where k_6 is the number of all orderings of type $\mathbb{Q}$.

Obviously, $k_1 + k_2 + \cdots + k_6 = m$.

[7] Let $\Gamma_s^{F_i}(x)$ be the formula which defines the sth element of the ith finite ordering, $1 \leq s \leq n_i$, $1 \leq i \leq k_1$. For example,

$$\Gamma_1^{F_i}(x) := F_i(x) \wedge \forall t[F_i(t) \wedge t < x \rightarrow E(t, x)].$$

[8] Let $\Gamma_s^{\omega_i}(x)$ be the formula which defines the sth element of the ith ordering of type ω, $1 \leq s < \omega$, $1 \leq i \leq k_2$.

[9] Let $\Gamma_s^{\omega_i^*}(x)$ be the formula which defines the sth element of the ith ordering of type ω^*, $1 \leq (-s) < \omega$, $1 \leq i \leq k_3$.

[10] Let $\Gamma_s^{(\omega+\omega^*)_i}(x)$ be the formula which defines the sth element of the ith ordering of type $\omega + \omega^*$, $s \in \mathbb{Z} \setminus \{0\}$, $1 \leq i \leq k_4$.

Let $s(x,y) := x < y \wedge \neg E(x,y) \wedge \forall t[x < t < y \rightarrow E(x,t) \vee E(t,y)]$. Obviously, $s(x,y) \Leftrightarrow$ "y is the immediate successor of x in M^+/E".

[11] Let for each $n < \omega$

$$\phi_n(x,y) := \exists z_1 \exists z_2 \cdots \exists z_{n-1}[s(x,z_1) \wedge s(z_1,z_2) \wedge \cdots \wedge s(z_{n-1},y)].$$

All formulas of points [1]–[11] including in addition atomic formulas we declare to be basic. It can now be shown by standard arguments that $Th(M^+)$ admits quantifier elimination relative to these basic formulas. Then it is not difficult to see that for any formula $\phi(x,\bar{y})$, for any model N of $Th(M^+)$ and for all $\bar{a} \in N$ $\phi(N,\bar{a})$ is a union of finitely many convex sets, i.e. $Th(M^+)$ is a weakly o-minimal theory, and consequently it is quite o-minimal (since $p \perp^w q$ for any $A \subseteq M^+$ and any non-algebraic distinct $p,q \in S_1(A)$).

Proposition 2.2. *Let M be an $\aleph_0$-categorical quite o-minimal structure non-having distinguished constants, M^+ be a linearly ordered disjoint E-combination of ω copies of M. Then M^+ is an $\aleph_0$-categorical quite o-minimal structure iff M is 1-indiscernible and M^+/E contains only finitely many elements having an immediate predecessor or an immediate successor.*

Proof of Proposition 2.2. ($\Rightarrow$) Then $Th(M^+)$ is quite o-minimal, and by Theorem 2.1 M is 1-indiscernible.

If we assume that M^+/E contains infinitely many elements having an immediate predecessor or an immediate successor then consider the following formula:

$$\phi(x) := \forall u(E(x,u) \rightarrow \exists y[\neg E(x,y) \wedge x < y \wedge \forall z(x < z < y \rightarrow$$

$$E(x,z) \vee E(z,y))]) \vee \forall u(E(x,u) \rightarrow \exists t[\neg E(x,t) \wedge t < x \wedge$$

$$\forall z(t < z < x \rightarrow E(t,z) \vee E(z,x))]).$$

By weak o-minimality of M^+ the set $\phi(M^+)$ is a union of finitely many convex sets. Consequently there is at least one convex part of $\phi(M^+)$ containing an infinite discretely ordered chain of E-classes. This contradicts $\aleph_0$-categoricity of M^+.

($\Leftarrow$) By proof of Theorem 2.1, M^+ is quite o-minimal. And since M^+/E contains only finitely many elements having an immediate predecessor or an immediate successor, we have that M^+ is $\aleph_0$-categorical.

Corollary 2.1. *Let M be an $\aleph_0$-categorical 1-indiscernible quite o-minimal structure non-having distinguished constants, M^+ be a linearly ordered disjoint E-combination of ω copies of M. If $Th(M^+)$ is quite o-minimal then either it is $\aleph_0$-categorical or has 2^ω countable models.*

Proof of Corollary 2.1. If M^+/E contains only finitely many elements having an immediate predecessor or an immediate successor then by Proposition 2.2, $Th(M^+)$ is $\aleph_0$-categorical. Suppose now that M^+/E contains infinitely many elements having an immediate predecessor or an immediate successor. Then by quite o-minimality of M^+ there is an infinite convex set containing an infinite discretely ordered chain of E-classes. Without loss of generality, suppose that it is an increasing chain. Take an arbitrary $a \in M^+$ belonging to this chain and consider the following formulas:

$$\phi_1(x,a) := E(a,x) \wedge a \leq x,$$

$$\phi_n(x,a) := \wedge_{i=1}^{n-1} \neg\phi_i(x,a) \wedge a < x \wedge \forall z[a < z < x \rightarrow \vee_{i=1}^{n-1} \phi_i(z,a) \vee E(z,x)],$$

for every $n \geq 2$.

Without loss of generality, we can also suppose that there exists an E-class E'' such that for every $n < \omega$, $\phi_n(M^+,a) < E''$. Obviously, there is some $b \in M^+$ with $E(M^+,b) = E''$.

Consider the following set of formulas:

$$C(x) := \{a < x\} \cup \{\neg\phi_n(x,a) \mid n \in \omega\} \cup \{\forall y[E(y,b) \rightarrow x < y]\}.$$

Obviously, $C(x)$ determines an $\{a,b\}$-definable cut in M^+. Since for every $n < \omega$ there exists an E-class E' such that $\phi_n(M,a) < E' < E''$, $C(x)$ can be realized by infinitely many E-classes without a greatest E-class, and between any discretely ordered chains of E-classes of an arbitrary finite or infinite length we can realize infinitely many densely ordered E-classes, whence $T(a,b)$ has 2^ω countable models (here $T(a,b) := Th(M^+) \cup \{\theta(a,b) \mid \theta(x,y)$ is an Σ-formula, $M_1^+ \models \theta(a,b)$ for some M_1^+ of $Th(M^+)$ with $\{a,b\} \subset M_1^+\}$). Then, obviously, $Th(M^+)$ has 2^ω countable models.

Note. Observe that in [13] M. Rubin proved the same result for complete theories of linear order with the language having a finite set of unary predicates. Here the language of $Th(M^+)$ may contain additionally finitely many binary predicates, and it contains at least one binary predicate (it is E).

Corollary 2.2. *There is no Ehrenfeucht quite o-minimal E-combination of countably many copies of an* $\aleph_0$*-categorical quite o-minimal structure non-having distinguished constants.*

3. Closures

As shown in[10,15] quite o-minimal theories with few countable models have the following basic characteristics in terms of non-algebraic 1-types: the number of independent types, convexity ranks of (independent) types, cardinalities of equivalence classes of dependent types.

Following these characteristics and the binarity of T by Theorem 5.4 of[10] we consider, without loss of generality, the language Σ consisting of unary predicates $P_{p,i}$, $P_{p,i+1} \subsetneq P_{p,i}$, for 1-types p: $p(x)$ is forced by $\{P_{p,i}(x) \mid i \in \omega\}$, and $P_{p,i}$ and $P_{q,i}$ are disjoint for distinct non-isolated p and q. Besides, Σ contains symbols $E_{p,j}$, $j \in \omega$, for equivalence relations on sets of realizations of p witnessing the convexity rank of p, as well as symbols $f_{p,q}$ for definable bijections between dependent types p and q.

We will permit some symbols to be interpreted by the empty set and if they interpreted as nonempty then their descriptions are consistent for the whole family. For any $E_{p,j+1}$ we assume that it is either empty or its equivalence classes divide $E_{p,j}$-classes on infinitely many convex $E_{p,j+1}$-classes, increasing the convexity rank of p under transformation $E_{p,j+1}$ (and all $E_{p,k}$, $k \leq j$) from the empty relation to an equivalence relation.

If T is $\aleph_0$-categorical then we can vary the convexity ranks of isolated formulas and links $f_{p,q}$ between types having the same convexity rank. Thus, for a family $\mathcal{T}$ with fixed 1-types new theories in the E-closure are exhausted by possibilities to increase the convexity rank till infinite one, (in)dependently for isolated types of T.

Having n independent non-algebraic 1-types p_i each type can obtain an infinite convexity rank. We can achieve these infinities by limits for tuples $(m_1, \ldots, m_n)$, where $m_i = RC(p_i)$ in a theory $T \in \mathcal{T}$.

Consider the tuples $\overline{m} = (m_1, \ldots, m_n) \in \omega^n$ characterizing theories in $\mathcal{T}$.

Clearly, $T \in \mathrm{Cl}_E(\mathcal{T})$ if and only if the tuple $(\mu_1, \ldots, \mu_n)$ for the convexity ranks of T is a limit for some convergent sequence $(\overline{m}^k)_{k\in\omega}$ of tuples $\overline{m}^k$ for theories in $\mathcal{T}$. Moreover, if $T \in \mathrm{Cl}_E(\mathcal{T}) \setminus \mathcal{T}$ then the sequence $(\overline{m}^k)_{k\in\omega}$ can be chosen *special*, i.e., without repetitions, with fixed finite coordinates, and with strictly increasing i-th coordinates for $\mu_i = \omega$.

Indeed, if $T \in \mathrm{Cl}_E(\mathcal{T}) \setminus \mathcal{T}$, with a tuple $\overline{\mu} = (\mu_1, \ldots, \mu_n)$, then T is approximated by a sequence of theories $T_k \in \mathcal{T}$, $k \in \omega$, with the corresponding tuples $\overline{m}^k \neq \overline{\mu}$. Since there are infinitely many distinct $\overline{m}^k$, we can choose a subsequence (T_{k_r}) such that the tuples $\overline{m}^{k_r}$ are pairwise distinct. Besides, if some coordinate μ_i is finite then infinitely many $\overline{m}^{k_r}$ have $m_i^{k_r} = \mu_i$. So $(\overline{m}^{k_r})_{r\in\omega}$ has a subsequence $(\overline{m}^{k_{r_s}})_{s\in\omega}$ with $m_i^{k_{r_s}} = \mu_i$. Now having infinite coordinates μ_i, $(\overline{m}^{k_{r_s}})_{s\in\omega}$ has a special subsequence $(\overline{m}^{k_{r_{s_t}}})_{t\in\omega}$, where $m_i^{k_{r_{s_t}}}$ are strictly increasing.

Special sequences $(\overline{l}^k)_{k\in\omega}$ and $(\overline{m}^k)_{k\in\omega}$ are *equivalent*, written $(\overline{l}^k) \sim (\overline{m}^k)$, if they have same fixed coordinates.

Lemma 3.1. *For a set $\mathcal{T}$ of $\aleph_0$-categorical quite o-minimal theories the following holds:*

(1) *$\sim$ is an equivalence relation on a set of special sequences.*

(2) *Special sequences $(\overline{l}^k)_{k\in\omega}$ and $(\overline{m}^k)_{k\in\omega}$ produce a common theory with a tuple $\overline{\mu} = (\mu_1, \ldots, \mu_n)$ if and only if $(\overline{l}^k) \sim (\overline{m}^k)$.*

(3) *Any special sequence is equivalent to each its proper subsequence.*

Proof. (1) Clearly, by the definition, that $\sim$ is reflexive and symmetric. Now if we assume that $(\overline{l}^k) \sim (\overline{m}^k)$ and $(\overline{m}^k) \sim (\overline{n}^k)$ then, since both $\overline{l}^k$ and $\overline{m}^k$ as well as $\overline{m}^k$ and $\overline{n}^k$ have the same fixed coordinates, $\overline{l}^k$ and $\overline{n}^k$ have the same fixed coordinates, too, i.e., $(\overline{l}^k) \sim (\overline{n}^k)$.

(2) If $(\overline{l}^k)_{k\in\omega}$ and $(\overline{m}^k)_{k\in\omega}$ produce a common theory with a tuple $\overline{\mu} = (\mu_1, \ldots, \mu_n)$ then by speciality of $(\overline{l}^k)_{k\in\omega}$ and $(\overline{m}^k)_{k\in\omega}$, the same coordinates are either fixed or strictly increasing. Therefore, $(\overline{l}^k) \sim (\overline{m}^k)$. Conversely, if $(\overline{l}^k) \sim (\overline{m}^k)$ then again by speciality of $(\overline{l}^k)_{k\in\omega}$ and $(\overline{m}^k)_{k\in\omega}$, the same coordinates are either fixed or strictly increasing, producing a common theory with a tuple $\overline{\mu} = (\mu_1, \ldots, \mu_n)$.

(3) Consider a special sequence $(\overline{l}^k)$. By the definition each coordinates either are fixed or strictly increasing. Taking an arbitrary subsequence $(\overline{m}^{k_s})$ we again obtain either fixed or strictly increasing coordinates, i.e., $(\overline{m}^{k_s})$ is special.

Proposition 3.1. *For an E-combination T_E of $\aleph_0$-categorical quite o-minimal theories forming a set $\mathcal{T}$ the following conditions are equivalent:*

(1) *e-Sp$(T_E) > 0$;*

(2) *$\mathcal{T}$ is infinite;*

(3) *$\mathcal{T}$ has a special sequence $(\overline{m}^k)_{k\in\omega}$.*

Proof. (1) $\Rightarrow$ (2). If $\mathcal{T}$ is finite then $\mathcal{T}$ is E-closed consisting of finitely many isolated theories whose all parameters μ_i are finite, whence e-Sp$(T_E) = 0$.

(2) $\Rightarrow$ (3). If $\mathcal{T}$ is infinite then some theories in $\mathcal{T}$ produce a special sequence $(\overline{m}^k)_{k\in\omega}$, since tuples $\overline{m}$ for theories $T \in \mathcal{T}$ have some unbounded coordinates m_i.

(3) $\Rightarrow$ (1). If $\mathcal{T}$ has a special sequence $(\overline{m}^k)_{k\in\omega}$, then its limit $\overline{\mu} = (\mu_1, \ldots, \mu_n)$ codes a theory $T \in \mathrm{Cl}_E(\mathcal{T}) \setminus \mathcal{T}$, with a type having infinite convexity rank. Therefore e-Sp$(T_E) > 0$.

Proposition 3.2. *For an E-combination T_E of $\aleph_0$-categorical quite o-minimal theories forming a set $\mathcal{T}$,* e-Sp(T_E) *equals the number of non-equivalent special sequences $(\overline{m}^k)_{k\in\omega}$ for $\mathcal{T}$. In particular,* e-Sp(T_E) *is finite if and only if $\mathcal{T}$ has only finitely many non-equivalent special sequences $(\overline{m}^k)_{k\in\omega}$.*

Proof follows immediately from Lemma 3.1 (2).

As already shown, obtaining E-closures for a fixed family of isolated 1-types $p_1, \ldots, p_n$, it suffices to consider special sequences, i.e., sequences of tuples $(m_1, \ldots, m_n)$ which are monotonic under all coordinates and strictly monotonic under some ones. In such a case we can vary sequences with infinite limits under some coordinates.

If for some theories $T_k \in \mathcal{T}$, $k \in \omega$, their tuples $(m_1^k, \ldots, m_n^k)$ have unbounded m_i^k with fixed m_j^k for $j \neq i$, then there is a theory $T \in \mathrm{Cl}_E(\mathcal{T})$ with infinite m_i and $m_j = m_j^k$ for $j \neq i$. Varying finite possibilities for m_j^k we can obtain $\lambda \in \omega \cup \{\omega\}$ theories in $\mathrm{Cl}_E(\mathcal{T}) \setminus \mathcal{T}$. Here, λ is finite if there are finitely many m_j^k, and $\lambda = \omega$ otherwise.

Thus we have the following:

Proposition 3.3. *For any $\lambda \in \omega + 1$ there is a family $\mathcal{T}$ of $\aleph_0$-categorical quite o-minimal theories such that all new theories in $\mathrm{Cl}_E(\mathcal{T}) \setminus \mathcal{T}$ are not $\aleph_0$-categorical and for the E-combination T_E of theories in $\mathcal{T}$, $|\mathrm{Cl}_E(\mathcal{T}) \setminus \mathcal{T}| =$ e-*Sp$(T_E) = \lambda$.

Theorem 3.1. *Families $\mathcal{T}$ of $\aleph_0$-categorical quite o-minimal theories produce E-combinations T_E with* e-Sp$(T_E) \in \omega + 1$. *All values in $\omega + 1$ are realized as e-spectra for some families of $\aleph_0$-categorical quite o-minimal theories.*

Proof. Since there are countably many possibilities for the tuples $(\lambda_1, \ldots, \lambda_n)$ of the convexity ranks for the theories in $\mathrm{Cl}_E(\mathcal{T})$ we obtain e-Sp$(T_E) \in \omega + 1$. Applying Proposition 3.3 we have all required values in $\omega+1$ as e-spectra for some families of $\aleph_0$-categorical quite o-minimal theories.

Dividing unary predicates $P_{p,i}$ by new, at most countably many, convex unary predicates we can obtain at most 2^ω non-isolated types. So, taking a family $\mathcal{T}$ of $\aleph_0$-categorical quite o-minimal theories allowing to vary (non)empty predicates $P_{p,i}$, we produce values $\lambda \in \omega \cup \{\omega, 2^\omega\}$ for e-spectra. Thus we have the following:

Proposition 3.4. *For any $\lambda \in \omega \cup \{\omega, 2^\omega\}$ there is a family $\mathcal{T}$ of $\aleph_0$-categorical quite o-minimal theories such that $|\mathrm{Cl}_E(\mathcal{T}) \setminus \mathcal{T}| = e\text{-Sp}(T_E) = \lambda$, witnessed by dividing of convex unary predicates, for the corresponding E-combination T_E of the theories in $\mathcal{T}$.*

Note. If we combine both the divisions of convex unary predicates $P_{p,i}$ and increases of the convexity ranks, we again obtain, applying Theorem 3.1 and Proposition 3.4 for any $\lambda \in \omega \cup \{\omega, 2^\omega\}$, a family $\mathcal{T}$ of $\aleph_0$-categorical quite o-minimal theories such that $|\mathrm{Cl}_E(\mathcal{T}) \setminus \mathcal{T}| = e\text{-Sp}(T_E) = \lambda$.

Note. Introducing $f_{p,q}$-links between 1-types p and q, we can regulate the number of independent types and the cardinalities for the classes of pairwise dependent types. Therefore, for instance, having countably many independent 1-types, expansions by $f_{p,q}$-links allow to produce a given number $\lambda \in \omega+1$ of equivalence classes of non-isolated 1-types such that each class consists of given (in)finitely many pairwise dependent types whereas types in distinct classes are independent. Thus, these tools produce a reformulation of Theorem 3.1 in the extended language.

Acknowledgments

This research has been funded by Science Committee of Ministry of Science and Higher Education of the Republic of Kazakhstan (Grant No. BR20281002), and in the framework of the State Contract of the Sobolev Institute of Mathematics, Project No. FWNF-2022-0012.

References

1. S.V. Sudoplatov, Combinations of structures, The Bulletin of Irkutsk State University. Series Mathematics, vol. 24 (2018), pp. 82–101.
2. S.V. Sudoplatov, Closures and generating sets related to combinations of structures, The Bulletin of Irkutsk State University. Series Mathematics, vol. 16 (2016), pp. 131–144.

3. S.V. Sudoplatov, Families of language uniform theories and their generating sets, The Bulletin of Irkutsk State University. Series Mathematics, vol. 17 (2016), pp. 62–76.
4. S.V. Sudoplatov, On semilattices and lattices for families of theories, Siberian Electronic Mathematical Reports, vol. 14 (2017), pp. 980–985.
5. S.V. Sudoplatov, Combinations related to classes of finite and countably categorical structures and their theories, Siberian Electronic Mathematical Reports, vol. 14 (2017), pp. 135–150.
6. S.V. Sudoplatov, Relative e-spectra, relative closures, and semilattices for families of theories, Siberian Electronic Mathematical Reports, vol. 14 (2017), pp. 296–307.
7. In.I. Pavlyuk, S.V. Sudoplatov, Families of theories of Abelian groups and their closures, Bulletin of Karaganda University. Mathematics, vol. 92, No. 4 (2018), pp. 72–78.
8. B.Sh. Kulpeshov, Convexity rank and orthogonality in weakly o-minimal theories, News of National Academy of Sciences of the Republic of Kazakhstan, series physics-mathematics, 227 (2003), pp. 26–31.
9. B.Sh. Kulpeshov, Weakly o-minimal structures and some of their properties, *The Journal of Symbolic Logic*, 63 (1998), pp. 1511–1528.
10. B.Sh. Kulpeshov, S.V. Sudoplatov, Vaught's conjecture for quite o-minimal theories,*Annals of Pure and Applied Logic*, 2017, vol. 168, issue 1,pp. 129–149.
11. L.L. Mayer, Vaught's conjecture for o-minimal theories, *The Journal of Symbolic Logic*, 53 (1988), pp. 146–159.
12. B.Sh. Kulpeshov, Countably categorical quite o-minimal theories, Journal of Mathematical Sciences, 188, issue 4 (2013), pp. 387–397.
13. M. Rubin, Theories of linear order, Israel Journal of Mathematics, 17, issue 4 (1974), pp. 392-443.
14. S.V. Sudoplatov, Classification of countable models of complete theories, Part 1. — Novosibirsk : Novosibirsk State Technical University Publishing House, 2018.
15. B.Sh. Kulpeshov, S.V. Sudoplatov, Linearly ordered theories which are nearly countably categorical, Mathematical Notes, 2017, volume 101, No. 3, pp. 475–483.

Geology of symmetric grounds

Toshimichi Usuba
Faculty of Fundamental Science and Engineering, Waseda University, Okubo 3-4-1, Shinjyuku, Tokyo, 169-8555 Japan
E-mail: usuba@waseda.jp

Let us say that a model of ZF is a symmetric ground if V is a symmetric extension of the model. In this paper, we investigate set-theoretic geology of symmetric grounds. Under a certain assumption, we show that all symmetric grounds of V are uniformly definable. We also show that if AC is forceable over V, then the symmetric grounds are downward directed.

Keywords: Axiom of choice, forcing method, set-theoretic geology, symmetric extension, symmetric ground.

1. Introduction

Let us say that a transitive model of ZF is a ZF-*ground*, or simply a *ground*, of V if V is a generic extension of the model. If a ground satisfies AC, the Axiom of Choice, then it is a ZFC-*ground*. In ZFC, Fuchs-Hamkins-Reitz[2] studied the structure of all ZFC-grounds of V, it is called *set-theoretic geology*. Their work was under AC, and Usuba[7] tried to study *set-theoretic geology without* AC; The universe V and each ground are not assumed to satisfy AC. These are attempts to investigate the nature of forcing method.

When we want to build choiceless models, the method of *symmetric submodel*, or *symmetric extension*, is a very powerful and flexible tool. For a generic extension $V[G]$ of V, a symmetric submodel of $V[G]$, or a symmetric extension of V, is realized as a submodel of the generic extension $V[G]$. The model $V[G]$ itself is a symmetric submodel of $V[G]$, so every generic extension is a symmetric extension. Moreover, symmetric extensions have many properties which are parallel to generic extensions, such as forcing relation and forcing theorem.

We want to say that a model W of ZF is a *symmetric ground* of V if V is a symmetric extension of W, that is, V is a symmetric submodel of *some* generic extension of W. Now we can expect to extend standard set-theoretic geology to one which treats symmetric extensions and symmetric grounds. If V is a symmetric extension of some model, then V could be a choiceless model. So our base theory should be ZF. On the other hand, our definition of symmetric grounds causes a problem: What is *some* generic extension of W? It would be possible that there is a W-generic G which is *not* living in V and in any generic extension of V, but V is realized as a symmetric submodel of $W[G]$. Hence, in V, it is not clear if we can describe that "W is a symmetric ground of V" and develop geology of symmetric grounds.

For this problem, using Grigorieff's work[3], we prove that the statement "W is a symmetric ground of V" is actually describable in V by a certain first order formula of the extended language $\{\in, W\}$. We also prove that, under a certain assumption, all symmetric grounds are uniformly definable by a first order formula of set-theory. Let CLS denote the assertion that there is a proper class of Löwenheim-Skolem cardinals (see Section 3).

Theorem 1.1 (in ZF). *Suppose* CLS. *Then there is a first order formula* $\varphi(x, y)$ *of set-theory such that:*

(1) For every set r, $\overline{W}_r = \{x \mid \varphi(x, r)\}$ *is a symmetric ground of* V *with* $r \in \overline{W}_r$.
(2) For every symmetric ground W *of* V, *there is* r *with* $W = \overline{W}_r$.

CLS follows from AC. Hence in ZFC, all symmetric grounds are uniformly definable. We also show that if V satisfies AC, then V is definable in any symmetric extension of V.

The above uniform definability of symmetric grounds allows us to investigate *set-theoretic geology of symmetric grounds* in ZF, which is a study of the structure of all symmetric grounds. This paper is a first step in the set-theoretic geology of symmetric grounds. In ZF, it is consistent that there are two grounds which have no common ground (Usuba[7]). We show that, if AC is forceable over V then all symmetric grounds are downward directed. Moreover the intersection of all symmetric grounds is a model of ZFC if AC is forceable over V.

2. Preliminaries

Throughout this paper, we do not assume AC unless otherwise specified. Forcing means a set forcing. A *class* means a second order object in the sense of Von Neumann-Bernays-Gödel set-theory NBG unless otherwise specified. We do not require that a class M is definable in V with some parameters, but we assume that V satisfies the comprehension and replacement scheme for the formulas of the language $\{\in, M\}$ (where we identify M as a unary predicate). Note that, if M is a definable class by a formula of the language $\{\in\}$, then V satisfies the comprehension and replacement scheme for the formulas of the language $\{\in, M\}$. Hence every definable class is a *class* in our sense. We also note that V is a class of any generic extension of V by the forcing theorem. Any theorems in this paper involving classes can be formalized in NBG, or some small extension of ZF.[a]

To treat classes in generic extensions, we extend the forcing language and relation as follows. Let M be a class. First, we extend the forcing language by adding the symbol $\check{M}$. For a poset $\mathbb{P}$, $p \in \mathbb{P}$, and a $\mathbb{P}$-name $\dot{x}$, we define $p \Vdash_{\mathbb{P}} \dot{x} \in \check{M}$ if the set $\{q \leq p \mid \text{there is } x \in M \text{ with } q \Vdash_{\mathbb{P}} x = \dot{x}\}$ is dense below p. $\check{M}$ can be treated as a class-sized canonical name, that is, $\check{M} = \{\langle \check{x}, p \rangle \mid p \in \mathbb{P}, x \in M\}$, where $\check{x}$ is the canonical name for x. By the standard way, we define $p \Vdash_{\mathbb{P}} \varphi$ for every formula φ of the extended forcing language. Since M is a class of V, we can easily check that the forcing theorem holds for the extended forcing language and relation, and $M = \{\dot{x}^G \mid p \Vdash_{\mathbb{P}} \dot{x} \in \check{M} \text{ for some } p \in G\}$. Using the extended forcing theorem, we have the following: If M is a class of V, then it is a class of any generic extension of V.

A model of ZF(C) means a transitive class model of ZF(C) containing all ordinals. It is known that a transitive class M containing all ordinals is a model of ZF if and only if M is closed under the Gödel operations and almost universal, that is, for every set $X \subseteq M$ there is a set $Y \in M$ with $X \subseteq Y$. So we can identify the sentence "M is a model of ZF" with the conjunction of the following sentences of the language $\{\in, M\}$:

(1) M is transitive and contains all ordinals.
(2) M is closed under the Gödel operations.
(3) M is almost universal.

A model W of ZF is a ZF-*ground*, or simply *ground*, of V if there are a poset $\mathbb{P} \in W$ and a $(W, \mathbb{P})$-generic G with $V = W[G]$. If a ZF-ground satisfies AC, then it is a ZFC-*ground*. **From now on, a ground means a ZF-ground unless otherwise specified.**

For a model M of ZF and an ordinal α, let $M_\alpha = M \cap V_\alpha$, the set of all $x \in M$ with rank less than α.

---

[a]For instance, we extend the language of set-theory by adding unary predicates $M, N, W, \ldots$ and let T be the theory ZF together with the comprehension and replacement scheme for the formulas of the language $\{\in, M, N, W, \ldots\}$.

Here we present a series of results by Grigorieff[3], which we will use frequently.

Theorem 2.1 (Grigorieff[3]). *For models M, N of* ZF*, if $M \subseteq N$, M is a ground of N, and N is a ground of V, then M is a ground of V.*

This means that the "is a ground of" relation is transitive in ZF.

For a model M of ZF and a set X, let $M(X) = \bigcup_{\alpha \in \mathrm{ON}} L(M_\alpha \cup \{X\})$. $M(X)$ is the minimal model of ZF with $M \subseteq M(X)$ and $X \in M(X)$. Note that $M(X)$ is also a class of V.

Theorem 2.2 (Theorem B in Grigorieff[3]). *Let $M \subseteq V$ be a ground of V. Let N be a model of* ZF *and suppose $M \subseteq N \subseteq V$. Then the following are equivalent:*

(1) V is a generic extension of N.
(2) N is of the form $M(X)$ for some $X \in N$.

For a class X, let HOD_X be the collection of all hereditarily definable sets with parameters from $\mathrm{ON} \cup X$. If M is a model of ZF, it is known that HOD_M is also a model of ZF with $M \subseteq \mathrm{HOD}_M$ [b], and HOD_M is a class of V (e.g., see Grigorieff[3]). We note that M is a class of HOD_M if M is a model of ZF.

Theorem 2.3 (9.3 Theorem 1 in Grigorieff[3]). *Let M be a model of* ZF *such that $V = M(X)$ for some $X \in V$. Then V is a ground of some generic extension of HOD_M, and $\mathrm{HOD}_M = M(Y)$ for some $Y \in V$ with $Y \subseteq M$.*

Remark 2.1. One may wonder what is "some generic extension of HOD_M". This theorem can be justified as follows: There is a poset $\mathbb{P}$ such that, in V, $\mathbb{P}$ forces that "$(\mathrm{HOD}_M)\check{}$ is a ground of the universe".

For a set S, let $\mathrm{Col}(S)$ be the poset consists of all finite partial functions from ω to S ordered by reverse inclusion. $\mathrm{Col}(S)$ is weakly homogeneous, and if S is ordinal definable then so is $\mathrm{Col}(S)$.

Theorem 2.4 (4.9 Theorem 1 in Grigorieff[3]). *Let $\mathbb{P}$ be a poset, and G be $(V, \mathbb{P})$-generic. Let α be a limit ordinal with $\alpha > \mathrm{rank}(\mathbb{P}) \cdot \omega$. Let H be $(V[G], \mathrm{Col}(V[G]_\alpha))$-generic. Then there is a $(V, \mathrm{Col}(V_\alpha))$-generic H' with $V[G][H] = V[H']$.*

---

[b]Note that HOD_M is not the relativisation of HOD to M.

The following is just rephrasing of this theorem:

Lemma 2.1. *Let M, N be grounds of V, and α a sufficiently large limit ordinal. Then whenever G is $(V, \mathrm{Col}(V_\alpha))$-generic, there are an $(M, \mathrm{Col}(M_\alpha))$-generic H and an $(N, \mathrm{Col}(N_\alpha))$-generic H' with $V[G] = M[H] = N[H']$.*

We recall some definitions and facts about symmetric submodels and extensions. See Jech[4] for more details. Let $\mathbb{P}$ be a poset, and $\mathrm{Auto}(\mathbb{P})$ the group of the automorphisms on $\mathbb{P}$. For $\pi \in \mathrm{Auto}(\mathbb{P})$, π can be extended to the (class) map from the $\mathbb{P}$-names to $\mathbb{P}$-names canonically. Let $\mathcal{G}$ be a subgroup of $\mathrm{Auto}(\mathbb{P})$. A non-empty family $\mathcal{F}$ of subgroups of $\mathcal{G}$ is a *normal filter* on $\mathcal{G}$ if:

(1) $H, H' \in \mathcal{F} \Rightarrow H \cap H' \in \mathcal{F}$.
(2) $H \in \mathcal{F}$, H' is a subgroup of $\mathcal{G}$ with $H \subseteq H'$ then $H' \in \mathcal{F}$.
(3) For $H \in \mathcal{F}$ and $\pi \in \mathcal{G}$, we have $\pi^{-1} H \pi \in \mathcal{F}$.

A triple $\langle \mathbb{P}, \mathcal{G}, \mathcal{F} \rangle$ is a *symmetric system* if $\mathcal{G}$ is a subgroup of $\mathrm{Auto}(\mathbb{P})$ and $\mathcal{F}$ is a normal filter on $\mathcal{G}$. A $\mathbb{P}$-name $\dot{x}$ is *$\mathcal{F}$-symmetric* if $\{\pi \in \mathcal{G} \mid \pi(\dot{x}) = \dot{x}\} \in \mathcal{F}$. Let $\mathrm{HS}_\mathcal{F}$ be the class of all hereditarily $\mathcal{F}$-symmetric names.

If G is $(V, \mathbb{P})$-generic, then $\mathrm{HS}_\mathcal{F}^G = \{\dot{x}^G \mid \dot{x} \in \mathrm{HS}_\mathcal{F}\}$ is a transitive model of ZF with $V \subseteq \mathrm{HS}_\mathcal{F}^G$. $\mathrm{HS}_\mathcal{F}^G$ is called a *symmetric submodel of* $V[G]$, or a *symmetric extension of* V. If we want to specify a poset $\mathbb{P}$, we say that $\mathrm{HS}_\mathcal{F}^G$ is a symmetric extension of V via a poset $\mathbb{P}$, or a symmetric submodel of $V[G]$ via a poset $\mathbb{P}$. We note that $\mathrm{HS}_\mathcal{F}^G$ is a class of $V[G]$, and V is a class of $\mathrm{HS}_\mathcal{F}^G$.

Remark 2.2. Let M be a model of ZF, $V[G]$ a generic extension of V via a poset $\mathbb{P}$, and $V \subseteq M \subseteq V[G]$. One may expect that if M is a symmetric extension of V, then M is a symmetric submodel of $V[G]$ via a poset $\mathbb{P}$. However this is not correct. This follows from the construction of the Bristol model (Karagila[5]). The Bristol model M is an intermediate model between L and the Cohen forcing extension $L[c]$, and M cannot be of the form $L(X)$ for some set X. M is of the form $\bigcup_{\alpha \in \mathrm{ON}} L(M_\alpha)$, and each $L(M_\alpha)$ is a symmetric extension of L, but there is some β such that for every $\alpha \geq \beta$, $L(M_\alpha)$ is not a symmetric submodel of $L[c]$ via the Cohen forcing notion.

The following theorem tells us when $M \subseteq V[G]$ is a symmetric submodel of $V[G]$.

Theorem 2.5 (Theorem C in Grigorieff[3]). *Let $M \subseteq V$ be a model of* ZF*, $\mathbb{P} \in M$ a poset, and G be $(V, \mathbb{P})$-generic and suppose $V = M[G]$. Let N be a model of* ZF *with $M \subseteq N \subseteq V$. Then N is a symmetric submodel of V via $\mathrm{ro}(\mathbb{P})^M$ if and only if N is of the form $\mathrm{HOD}_{M(X)}$ for some $X \in N$, where $\mathrm{ro}(\mathbb{P})$ is the completion of $\mathbb{P}$.*[c]

3. Choiceless geology

In this section, we recall some definitions and facts about set-theoretic geology without AC. See Usuba[7] for more information. For Löwenheim-Skolem cardinals below, see also Usuba[8].

We use the notion of *Löwenheim-Skolem cardinal*, which was introduced in Usuba[7]. The notion of Löwenheim-Skolem cardinal corresponds to the downward Löwenheim-Skolem theorem in the context of ZFC. The downward Löwenheim-Skolem theorem is not provable from ZF, but under the existence of Löwenheim-Skolem cardinals, we can take *small* elementary submodels of V_α.

Definition 3.1. An uncountable cardinal κ is a *Löwenheim-Skolem cardinal* if for every $\alpha \geq \kappa$, $x \in V_\alpha$, and $\gamma < \kappa$, there are $\beta \geq \alpha$ and $X \prec V_\beta$ satisfying the following:

(1) $x \in X$ and $V_\gamma \subseteq X$.
(2) ${}^{V_\gamma}(X \cap V_\alpha) \subseteq X$.
(3) The transitive collapse of X belongs to V_κ.

Note that every limit of Löwenheim-Skolem cardinals is a Löwenheim-Skolem cardinal as well.

Theorem 3.1 (Usuba[7]). *Suppose κ is a limit of Löwenheim-Skolem cardinals. Then for every poset $\mathbb{P} \in V_\kappa$, $\mathbb{P}$ forces that "κ is a Löwenheim-Skolem cardinal". Conversely, if $\kappa < \lambda$ are cardinals, $\mathbb{P} \in V_\kappa$ a poset, and $\Vdash_{\mathbb{P}}$ "κ, λ are Löwenheim-Skolem ", then λ is Löwenheim-Skolem in V.*

Definition 3.2. For a set X, the *norm* of X, denoted by $\|X\|$, is the least ordinal α such that there is a surjection from V_α onto X.

Definition 3.3. Let $M \subseteq V$ be a model of (a sufficiently large fragment of) ZF, and α an ordinal.

---

[c]We can take the completion of $\mathbb{P}$ without AC, see Grigorieff[3].

(1) M has the *α-norm covering property* for V if for every set $X \subseteq M$, if $\|X\| < \alpha$ then there is $Y \in M$ such that $X \subseteq Y$ and $\|Y\|^M < \alpha$.
(2) M has the *α-norm approximation property* for V if for every set $Y \subseteq M$, if $Y \cap a \in M$ for every $a \in M$ with $\|a\|^M < \alpha$ then $Y \in M$.

Theorem 3.2 (Usuba[7]). *Suppose κ is a Löwenheim-Skolem cardinal and $\alpha < \kappa$. For all models M, N of (a sufficiently large fragment of) ZF, if M and N have the α-norm covering and the α-norm approximation properties for V and $M_\kappa = N_\kappa$, then $M = N$.*

Theorem 3.3 (Usuba[7]). *Let κ be a Löwenheim-Skolem cardinal, and $W \subseteq V$ a model of ZF. If there are a poset $\mathbb{P} \in W_\kappa$ and a $(W, \mathbb{P})$-generic G with $V = W[G]$, then W has the κ-norm covering and the κ-norm approximation properties for V.*

Let us say that a model M of ZF is a *pseudo-ground* if M has the α-norm covering and the α-norm approximation properties for V for some α.

The following is immediate:

Lemma 3.1. *Let $M \subseteq N \subseteq V$ be models of ZF, and α an ordinal.*

(1) If M has the α-norm covering and the α-norm approximation properties for N, and N for V, then M has the α-norm covering and the α-norm approximation properties for V.
(2) If M has the α-norm covering and the α-norm approximation properties for V, then M has the α-norm covering and the α-norm approximation properties for N.

Definition 3.4. Let CLS denote the assertion that "there is a proper class of Löwenheim-Skolem cardinals".

By Theorem 3.1, CLS is absolute between V, all generic extensions, and all grounds of V.

Theorem 3.4 (Usuba[7]). *Suppose CLS. Then all pseudo-grounds are uniformly definable: There is a first order formula $\varphi'(x, y)$ of set-theory such that:*

(1) For every $r \in V$, the class $W'_r = \{x \mid \varphi'(x, r)\}$ is a pseudo-ground of V with $r \in W'_r$.
(2) For every pseudo-ground W of ZF, there is r with $W = W'_r$.

We sketch the proof since we will need to know how to define φ' and W'_r later.

Sketch of the proof. For $r \in V$, suppose r fulfills the following properties:

(1) r is of the form $\langle X, \kappa, \alpha \rangle$ where κ is a Löwenheim-Skolem cardinal, $\alpha < \kappa$, and X is a transitive set with $X \cap \mathrm{ON} = \kappa$.
(2) For each cardinal $\lambda > \kappa$, if V_λ is a model of a sufficiently large fragment of ZF, then there is a unique transitive model $X^{r,\lambda}$ of a sufficiently large fragment of ZF such that $X^{r,\lambda} \cap \mathrm{ON} = \lambda$, $(X^{r,\lambda})_\kappa = X$, and $X^{r,\lambda}$ has the α-norm covering and the α-norm approximation properties for V_λ.

Then let $W'_r = \bigcup_{\lambda > \kappa} W^{r,\lambda}$. Otherwise, let $W'_r = V$. By Theorem 3.2, one can check that the collection $\{W'_r \mid r \in V\}$ is as required. $\square$

Using this, we have the uniform definability of all grounds under CLS.

Theorem 3.5 (Usuba[7]). *Suppose* CLS. *Then there is a first order formula $\varphi(x, y)$ of set-theory such that:*

(1) For every $r \in V$, the class $W_r = \{x \mid \varphi(x, r)\}$ is a ground of V with $r \in W_r$.
(2) For every ground W of ZF, *there is r with $W = W_r$.*

Sketch of the proof. Let $\{W'_r \mid r \in V\}$ be the definable collection of pseudo-grounds of V as in Theorem 3.4. If W'_r is a ground of V then put $W_r = W'_r$, and $W_r = V$ otherwise. By Theorem 3.3, the definable collection $\{W_r \mid r \in V\}$ comprises all grounds of V. $\square$

Remark 3.1. In addition, we have the following by Theorem 3.1: Suppose CLS. If the formula φ defines all grounds as above, then, in all generic extensions, the formula φ defines all its grounds by the same way.

Lemma 3.2. *Suppose* CLS. *Then there is a first order formula $\psi(x, y, z)$ in the language of set-theory such that for every poset $\mathbb{P}$, $\mathbb{P}$-name $\dot{s}$, and a set t, $\Vdash_{\mathbb{P}}$ "$W_{\dot{s}}^{V[\dot{G}]} = \check{W}'_t$" if and only if $\psi(\mathbb{P}, \dot{s}, t)$ holds. Hence the statement $\Vdash_{\mathbb{P}}$ "$\check{W}'_t$ is a ground of the universe" is a first order assertion as $\exists \mathbb{P} \exists \dot{s} \psi(\mathbb{P}, \dot{s}, t)$.*

Proof. Let $\psi(\mathbb{P}, \dot{s}, t)$ be the following sentence: There are Löwenheim-Skolem cardinals $\kappa < \lambda$ which are limits of Löwenheim-Skolem cardinals and $\mathbb{P} \in V_\kappa$ such that:

(1) $\Vdash_{\mathbb{P}}$"$W_{\dot{s}}^{V[\dot{G}]}$ is a ground of the universe $V[\dot{G}]$ via a poset $\mathbb{Q} \in (W_{\dot{s}}^{V[\dot{G}]})_\kappa$".

(2) For every $p \in \mathbb{P}$, there is $q \leq p$ and a set X such that $q \Vdash_{\mathbb{P}}$ "$X = (W_{\dot{s}}^{V[\dot{G}]})_\lambda$", $X = (W'_u)_\lambda$ where $u = \langle X, \lambda, \kappa \rangle$, and $W'_u = W'_t$.

We see that this ψ works.

First, suppose $\Vdash_{\mathbb{P}}$ "$W_{\dot{s}}^{V[\dot{G}]} = \check{W}'_t$". By CLS, there are large Löwenheim-Skolem cardinals $\kappa < \lambda$ which are limits of Löwenheim-Skolem cardinals such that $\Vdash_{\mathbb{P}}$ "$W_{\dot{s}}^{V[\dot{G}]}$ is a ground of the universe $V[\dot{G}]$ via poset $\mathbb{Q} \in (W_{\dot{s}}^{V[\dot{G}]})_\kappa$". We see that κ and λ witness $\psi(\mathbb{P}, \dot{s}, t)$.

By Theorems 3.1 and 3.3, the following hold:

(1) $\Vdash_{\mathbb{P}}$ "κ and λ are Löwenheim-Skolem cardinals".
(2) $\Vdash_{\mathbb{P}}$ "$W_{\dot{s}}^{V[\dot{G}]}$ has the κ-norm covering and the κ-norm approximation properties for $V[\dot{G}]$".
(3) $\Vdash_{\mathbb{P}}$ "$\check{V}$ has the κ-norm covering and the κ-norm approximation properties for $V[\dot{G}]$".

Note that whenever G is $(V, \mathbb{P})$-generic and $s = \dot{s}^G$, we have $W'_t = W_s^{V[G]} \subseteq V \subseteq V[G]$. Since $W'_t = W_s^{V[G]}$ has the κ-norm covering and the κ-approximation properties for $V[G]$, we know that $W'_t = W_s^{V[G]}$ has the κ-norm covering and the κ-approximation properties for V by Lemma 3.1.

Now take $p \in \mathbb{P}$. Since $\Vdash_{\mathbb{P}}$ "$W_{\dot{s}}^{V[\dot{G}]} = \check{W}'_t \subseteq \check{V}$", we can choose $q \leq p$ and a set X such that $q \Vdash_{\mathbb{P}}$ "$X = (W_{\dot{s}}^{V[\dot{G}]})_\lambda$". Put $u = \langle X, \lambda, \kappa \rangle$. W'_t has the κ-norm covering and the κ-norm approximation properties for V and $X = (W'_t)_\lambda$. Hence $W'_t = W'_u$ by the definition of W'_u and Theorem 3.2.

For the converse, suppose $\psi(\mathbb{P}, \dot{s}, t)$ holds. Fix Löwenheim-Skolem cardinals $\kappa < \lambda$ witnessing $\psi(\mathbb{P}, \dot{s}, t)$. Fix $p \in \mathbb{P}$, and take $q \leq p$ and a set X such that $q \Vdash_x$ "$X = (W_{\dot{s}}^{V[\dot{G}]})_\lambda$". Let $u = \langle X, \lambda, \kappa \rangle$. We know that $W'_t = W'_u$, $(W'_u)_\lambda = X$, and $W'_t = W'_u$ has the κ-norm covering and the κ-norm approximation properties for V.

Now take a $(V, \mathbb{P})$-generic G. In $V[G]$, κ and λ remain Löwenheim-Skolem cardinals, and $W_s^{V[G]}$ has the κ-norm covering and the κ-norm approximation properties for $V[G]$. Moreover V has the κ-norm covering and the κ-norm approximation properties for $V[G]$. Then so does W'_u for $V[G]$ by Lemma 3.1. Applying Theorem 3.2, we have $W'_t = W'_u = W_s^{V[G]}$. □

Definition 3.5 (Blass[1]). The principle SVC (Small Violation of Choice) is the assertion that there is a set S such that for every set X, there is an ordinal α and a surjection from $S \times \alpha$ onto X.

Theorem 3.6 (Blass[1]). *The following are equivalent:*

(1) SVC *holds.*
(2) AC *is forceable, that is, there is a poset* $\mathbb{P}$ *which forces* AC.

By this theorem, we know that SVC is absolute between V, any ground, and any generic extension of V.

Theorem 3.7 (Usuba[7]). *Suppose* SVC *holds. Then* CLS *holds, in particular all grounds of* V *are uniformly definable.*

4. Symmetric grounds

In this section, we study a characterization of symmetric grounds without taking *some* generic extension, and we prove the uniform definability of symmetric grounds under CLS.

Lemma 4.1. *Let* M, N *be models of* ZF *and suppose* M *is a ground of* V, *say* $V = M[G]$. *If* $M \subseteq N \subseteq V$ *and* N *is a symmetric submodel of* $V = M[G]$, *then* N *is of the form* $M(X)$ *for some* X.

Proof. By Theorem 2.5, N is of the form $\mathrm{HOD}_{M(Y)}$ for some Y. Since $V = M[G] = M(Y)(G)$, we know that $N = \mathrm{HOD}_{M(Y)}$ is of the form $M(Y)(Z)$ for some Z with $Z \subseteq M(Y)$ by Theorem 2.3. Then $N = M(\{Y, Z\})$. □

Lemma 4.2. *Let* M, N *be models of* ZF *and suppose* M *is a ground of* V, *say* $V = M[G]$. *Suppose* $M \subseteq N \subseteq V$, M *is a class of* N, *and* N *is of the form* $M(X)$ *for some* X. *Then there is a large limit* α *such that whenever* G' *is* $(N, \mathrm{Col}(N_\alpha))$, *there is an* $(M, \mathrm{Col}(M_\alpha))$*-generic* H *such that* $N[G'] = M[H]$ *and* N *is a symmetric submodel of* $M[H]$.

Remark 4.1. As in Remark 2.1, the conclusion of this proposition can be read as follows: There is some α such that, in N, $\mathrm{Col}(N_\alpha)$ forces "there is an $(\check{M}, \mathrm{Col}(M_\alpha))$-generic H such that $\check{M}[H]$ is the universe, and $\check{N}$ is a symmetric submodel of $\check{M}[H]$".

Proof. By Theorem 2.2, $V = M[G]$ is a generic extension of N. Applying Lemma 2.1, there is a large limit α such that if G is $(V, \mathrm{Col}(V_\alpha))$-generic, there is an $(M, \mathrm{Col}(M_\alpha))$-generic H and an $(N, \mathrm{Col}(N_\alpha))$-generic H' with $V[G] = M[H] = N[H']$. Then, in N, $\mathrm{Col}(N_\alpha)$ forces that "there is an $(\check{M}, \mathrm{Col}(M_\alpha))$-generic H such that $\check{M}[H]$ is the universe". Hence whenever G' is $(N, \mathrm{Col}(N_\alpha))$-generic, there is an $(M, \mathrm{Col}(M_\alpha))$-generic H such that $N[G'] = M[H]$. We shall see that N is of the form $(\mathrm{HOD}_{M(X)})^{M[H]}$, hence N is a symmetric submodel of $M[H]$ by Theorem 2.5. Since $N = M(X)$, it is clear that $N \subseteq (\mathrm{HOD}_{M(X)})^{M[H]}$. Because $\mathrm{Col}(N_\alpha)$ is a weakly homogeneous ordinal definable poset in N, we have $(\mathrm{HOD}_{M(X)})^{M[H]} = (\mathrm{HOD}_N)^{N[G']} \subseteq N$. □

We have the following observation. Roughly speaking, it asserts that W is a symmetric ground of V if and only if W is a ground of *some* generic extension of V.

Proposition 4.1. *Let $M \subseteq N$ be models of* ZF, *and suppose M is a class of N.*

(1) Suppose M is a ground of V, say $V = M[G]$. If N is a symmetric submodel of $M[G] = V$, then there are a poset $\mathbb{P} \in N$ and $\mathbb{Q} \in M$ such that, in N, $\mathbb{P}$ forces "there is an $(\check{M}, \mathbb{Q})$-generic H such that $\check{M}[H]$ is the universe".

(2) If there are a poset $\mathbb{P} \in N$ and $\mathbb{Q} \in M$ such that, in N, $\mathbb{P}$ forces "there is an $(\check{M}, \mathbb{Q})$-generic H such that $\check{M}[H]$ is the universe", then N is a symmetric submodel of some generic extension of M.

Proof. (1). By Lemma 4.1, N is of the form $M(X)$ for some set $X \in N$. Then by Lemma 4.2, there is some large α such that, in N, $\mathrm{Col}(N_\alpha)$ forces that "there is an $(\check{M}, \mathrm{Col}(\check{M}_\alpha))$-generic H such that $\check{M}[H]$ is the universe".

(2). Take an $(N, \mathbb{P})$-generic G. In $N[G]$, there is $H \subseteq \mathbb{Q} \in M$ with $N[G] = M[H]$. $M \subseteq N \subseteq N[G] = M[H]$, hence N is of the form $M(X)$ for some $X \in M[H]$ by Theorem 2.2, and N is a symmetric submodel of some generic extension of $M[H]$ by Lemma 4.2. □

Using this proposition, we can obtain a formal definition of symmetric grounds.

Definition 4.1. Let W be a model of ZF. Let us say that W is a *symmetric ground* of V, or V is a *symmetric extension* of W, if there are posets $\mathbb{P} \in V$ and $\mathbb{Q} \in W$ such that $\mathbb{P}$ forces that "there is a $(\check{W}, \mathbb{Q})$-generic H such that $\check{W}[H]$ is the universe".

By Proposition 4.1, our notion of symmetric grounds and extensions coincide with the standard definition of symmetric submodels and extensions. Moreover our notion of symmetric extension is equivalent to quasi-generic extension in Grigorieff[3].

Remark 4.2. In Definition 4.1, V need not be a symmetric extension of W via a poset $\mathbb{Q}$.

Remark 4.3. If W is a symmetric ground of V, then $V = M(X)$ for some set X by Lemma 4.1. But the converse does not hold; For instance, if $0^\#$ exists, then $V = L(0^\#) = L[0^\#]$ is not a symmetric extension of L.

By Lemma 2.1, posets $\mathbb{P}$ and $\mathbb{Q}$ can be Levy collapsings.

Lemma 4.3. *Let W be a model of* ZF. *Then W is a symmetric ground of V if and only if for every sufficiently large limit ordinal α,* $\mathrm{Col}(V_\alpha)$ *forces "there is a $(\check{W}, \mathrm{Col}(\check{W}_\alpha))$-generic H such that $\check{W}[H]$ is the universe".*

Now we obtain a first order definition of symmetric grounds under CLS.

Theorem 4.1. *Suppose* CLS *(e.g.,* SVC *holds). Then there is a first order formula $\varphi(x, y)$ of set-theory such that:*

(1) For every set r, $\overline{W}_r = \{x \mid \varphi(x, r)\}$ is a symmetric ground of V with $r \in \overline{W}_r$.
(2) For every symmetric ground W of V, there is r with $W = \overline{W}_r$.

Proof. Let $\{W_r' \mid r \in V\}$ be the collection of all pseudo-grounds defined as in Theorem 3.4. Define $\overline{W}_r$ as follows: If there is a poset $\mathbb{P}$ such that $\Vdash_{\mathbb{P}}$"$\check{W}_r'$ is a ground of the universe", then $\overline{W}_r = W_r'$. Otherwise $\overline{W}_r = V$. By Lemma 3.2, $\{\overline{W}_r \mid r \in V\}$ is a first order definable collection. We check that $\{\overline{W}_r \mid r \in V\}$ is the collection of all symmetric grounds of V.

For $r \in V$, if $\overline{W}_r \neq V$ then $\overline{W}_r = W_r'$ and there is a poset $\mathbb{P}$ such that $\Vdash_{\mathbb{P}}$"$\check{W}_r'$ is a ground of the universe". Hence W_r' is a ground of a generic extension of V via $\mathbb{P}$, so $\overline{W}_r = W_r'$ is a symmetric ground of V. For the converse, suppose W is a symmetric ground of V. We can choose a generic extension $V[G]$ of V via poset $\mathbb{P}$ such that W is a ground of $V[G]$. We can take a large Löwenheim-Skolem cardinal κ such that κ is Löwenheim-Skolem in V[G], and W have the κ-norm covering and the κ-norm approximation properties for $V[G]$. Since $W \subseteq V \subseteq V[G]$, W has the κ-norm covering and the κ-norm approximation properties for V. Hence $W = W_r'$ for some $r \in V$, and in V we can choose a poset $\mathbb{Q}$ with $\Vdash_{\mathbb{Q}}$"$\check{W}_r'$ is a ground of the universe", so $W = \overline{W}_r$. □

Hence in ZFC, we can always define all symmetric grounds uniformly.

5. Some properties of symmetric extensions and grounds

In this section, we make some observations and prove some useful properties of symmetric extensions and grounds.

The following is partially proved by Grigorieff[3].

Lemma 5.1. *Let M and N be models of* ZF *with $M \subseteq N \subseteq V$. Then any two of the following conditions imply the third:*

(1) N is a symmetric extension of M.
(2) V is a symmetric extension of N.
(3) V is a symmetric extension of M.

Proof. (1)$\wedge$(2) $\Rightarrow$ (3). By Lemma 4.3, there is some large limit ordinal α such that:

- In N, $\mathrm{Col}(N_\alpha)$ forces that "there is an $(\check{M}, \mathrm{Col}(\check{M}_\alpha))$-generic H such that $\check{M}[H]$ is the universe".
- In V, $\mathrm{Col}(V_\alpha)$ forces that "there is an $(\check{N}, \mathrm{Col}(\check{N}_\alpha))$-generic H such that $\check{N}[H]$ is the universe".

Take a $(V, \mathrm{Col}(V_\alpha))$-generic G. Then there is an $(N, \mathrm{Col}(N_\alpha))$-generic H with $V[G] = N[H]$, and there is an $(M, \mathrm{Col}(M_\alpha))$-generic H' with $M[H'] = N[H]$. Hence M is a ground of the generic extension $V[G]$ of V, and M is a symmetric ground of V.

(1)$\wedge$(3) $\Rightarrow$ (2). Since V is a symmetric extension of M, there is a generic extension $M[G]$ of M such that $V \subseteq M[G]$ and $V = M(X)$ for some $X \in V$ by Lemma 4.1. By the same argument, $N = M(Y)$ for some $Y \in N$. We know $M \subseteq N = M(Y) \subseteq V = M(X) \subseteq M[G]$. Then $M[G]$ is a generic extension of $N = M(Y)$ by Theorem 2.2. We know $V = N(X)$, hence V is a symmetric extension of N by Lemma 4.2.

(2)$\wedge$(3) $\Rightarrow$ (1). By Lemma 4.3, there is some large α such that whenever G is $(V, \mathrm{Col}(V_\alpha))$-generic, we have that $V[G] = N[H] = M[H']$ for some $H \subseteq \mathrm{Col}(N_\alpha) \in N$ and $H' \subseteq \mathrm{Col}(M_\alpha) \in M$. Hence M is a ground of $N[H]$, so M is a symmetric ground of N. □

By this lemma, the "is a symmetric extension of" relation is transitive.

It is known that if $M \subseteq N \subseteq M[G]$ are models of ZFC and $M[G]$ is a generic extension of M, then M is a ground of N, and N is of $M[G]$ (compare Theorem 2.2). We prove a variant of this fact in our context.

Lemma 5.2. *Suppose M is a symmetric ground of V. Then for every model N of* ZF *with $M \subseteq N \subseteq V$, the following are equivalent:*

(1) N is a symmetric ground of V.
(2) N is of the form $M(X)$ for some set $X \in V$.
(3) N is a symmetric extension of M.

Proof. (1) $\iff$ (3) is immediate from Lemma 5.1.

(2) $\Rightarrow$ (1). Since M is a symmetric ground, there is a generic extension $V[G]$ of V such that M is a ground of $V[G]$. Then $N = M(X)$ is a ground of $V[G]$ by Theorem 2.2, so N is a symmetric ground of V.

(1) $\Rightarrow$ (2) is Lemma 4.1. □

It is also known that the Bristol model[5] is an intermediate model N between the constructible universe L and the Cohen forcing extension of L but N does not satisfy (1)–(3) of Lemma 5.2.

We can characterize SVC in terms of symmetric grounds. Note that under AC, all grounds are uniformly definable.

Theorem 5.1 (Usuba[6]). *Suppose V satisfies* AC, *and let $\{W_r \mid r \in V\}$ be the collection of all* ZFC*-grounds of V. Then for every set X, there is a set s such that $W_s \subseteq \bigcap_{r \in X} W_r$.*

Proposition 5.1. *The following are equivalent:*

(1) SVC *holds.*
(2) There is a symmetric ground satisfying AC.
(3) There is a symmetric ground W satisfying AC *and $V = W(X)$ for some set X.*

Proof. (1) $\Rightarrow$ (2). Take a poset $\mathbb{P}$ which forces AC. Take a $(V, \mathbb{P}\times\mathbb{P})$-generic $G \times H$. Both $V[G]$ and $V[H]$ are grounds of $V[G][H]$ and satisfy AC. By Theorem 5.1, there is a ground W of $V[G][H]$ such that W satisfies AC and $W \subseteq V[G] \cap V[H]$. Since $V[G] \cap V[H] = V$ (e.g., see Fuchs-Hamkins-Reitz[2]), we have $W \subseteq V \subseteq V[G][H]$. Then V is of the form $W(X)$ for some set X by Theorem 2.2. Now V is a symmetric extension of W by Lemma 5.2.

(2) $\Rightarrow$ (1). There is a generic extension $W[G]$ of W such that V is a ground of $W[G]$. $W[G]$ satisfies AC, so $W[G]$ witnesses that SVC holds in V.

(2) $\Rightarrow$ (3). If W is a symmetric ground satisfying AC, then W is a ground of some generic extension $V[G]$ of V. Applying Theorem 2.2, we have that $V = W(X)$ for some set $X \in V$.

(3) $\Rightarrow$ (2) is trivial. □

Next we prove the absoluteness of CLS and SVC between all symmetric grounds and symmetric extensions.

Proposition 5.2. *Let W be a symmetric ground of V. Then* CLS *and* SVC *are absolute between W and V.*

Proof. Since V is a symmetric extension of W, we have that V is a ground of some generic extension $W[G]$ of W. Since CLS and SVC are absolute between all grounds and generic extensions, it is also absolute between W, V, and $W[G]$. □

By the absoluteness of CLS and Theorem 4.1, we have:

Corollary 5.1. *Suppose* CLS *holds. Then V is definable in its symmetric extensions with parameters from V.*[d]

In particular, under AC, V is always definable in its symmetric extensions.

Remark 5.1. As the uniform definability of grounds, under CLS, if the formula φ defines all symmetric grounds as in Theorem 4.1, then in all symmetric grounds and symmetric extensions of V, φ also defines all its symmetric grounds. This follows from the absoluteness of CLS and the proof of Theorem 4.1.

Finally, we observe the structure of the symmetric models under AC.

Lemma 5.3. *Suppose V satisfies* AC. *Then every symmetric ground satisfying* AC *is a ground.*

Proof. Let W be a symmetric ground of V. Then W is a ground of some generic extension $V[G]$ of V. V, $V[G]$, and W are models of ZFC with $W \subseteq V \subseteq V[G]$. Then W is a ground of V. □

Corollary 5.2. *Suppose V satisfies* AC. *Then for every model M of* ZF, *the following are equivalent:*

(1) M is a symmetric ground of V.
(2) M is a ground of V.
(3) There is a ZFC-*ground W of V and a set X such that $M = W(X)$.*

Proof. (3) ⇒ (2). If $M = W(X)$ for some ground W of V, then M is a ground by Theorem 2.2.

(2) ⇒ (1) is trivial.

(1) ⇒ (3). Suppose M is a symmetric ground of V. Since V satisfies SVC trivially, SVC holds in M as well by Proposition 5.2. Then M has

[d] A. Karagila independently obtained this result assuming that V satisfies AC.

a symmetric ground W satisfying AC and a set X with $M = W(X)$ by Proposition 5.1. By Lemma 5.1, W is a symmetric ground of V, and in fact it is a ground by Lemma 5.3. □

We also note the following, which contrasts with Lemma 5.2.

Corollary 5.3. *Let M be a symmetric ground of V. Then for every model N of* ZF + SVC *with $M \subseteq N \subseteq V$, N is a symmetric ground of V, and N is a symmetric extension of M.*

Proof. By Proposition 5.1, N has a symmetric ground W satisfying AC and N is of the form $W(X)$. Since M is a symmetric ground of V, there is a set Y with $V = M(Y)$. Then clearly $V = W(\{X, Y\})$, and SVC holds in V. Force with $\mathrm{Col}(\mathrm{trcl}(\{X, Y\}))$ over V. This produces a generic extension $V[G]$ of V such that $V[G]$ which satisfies AC. Moreover SVC holds in M as well by Proposition 5.2. Thus there is a symmetric ground U of M such that U satisfies AC by Proposition 5.1 again. Then U is a symmetric ground of $V[G]$, hence is a ground of $V[G]$ by Lemma 5.3. Now $U \subseteq W \subseteq V[G]$ and $V[G]$ is a generic extension of U, so W is a ground of $V[G]$. Therefore $N = W(X)$ is a symmetric ground of $V[G]$, and so N is a symmetric ground of V. In addition N is a symmetric extension of M by Lemma 5.2. □

6. The downward directedness of symmetric grounds

In ZFC, the collection of ZFC-grounds is downward directed (Theorem 5.1). Unlike in the ZFC-context, under ZF+SVC, it is possible that V has two grounds which have no common ground (see Usuba[7]). On the other hand, we prove that under SVC, all symmetric grounds are downward directed.

Proposition 6.1. *Suppose* SVC *holds (hence all symmetric grounds are uniformly definable), and let $\{\overline{W}_r \mid r \in V\}$ be the collection of all symmetric grounds. Then for every set X, there is a symmetric ground W of V such that $W \subseteq \bigcap_{r\in X} \overline{W}_r$, and W is a symmetric ground of each $\overline{W}_r$ for $r \in X$.*

Proof. By Corollary 5.2, SVC holds in $\overline{W}_r$ for every $r \in X$. By Proposition 5.1, each $\overline{W}_r$ has a symmetric ground satisfying AC. Every symmetric ground of $\overline{W}_r$ is a symmetric ground of V as well by Lemma 5.1. Hence we can find a set Y such that for each $s \in Y$, $\overline{W}_s$ satisfies AC, and for every $r \in X$ there is $s \in Y$ with $\overline{W}_s \subseteq \overline{W}_r$.

Take a generic extension $V[G]$ of V such that AC holds in $V[G]$. By Lemma 5.3, for each $s \in Y$, $\overline{W}_s$ is a ground of $V[G]$. By Theorem 5.1, there is a ground W of $V[G]$ such that W satisfies AC and $W \subseteq \overline{W}_s$ for every $s \in Y$. Now, for each $s \in Y$, we have $W \subseteq \overline{W}_s \subseteq V \subseteq V[G]$. Since $V[G]$ is a generic extension of V and W, we know that W is a symmetric ground of V. By the choice of Y, we have that $W \subseteq \bigcap_{r \in X} \overline{W}_r$. In addition, by Lemma 5.1, W is a symmetric ground of $\overline{W}_r$ for every $r \in X$. □

7. Mantles

If all grounds are uniformly definable as in Theorem 3.5, then we can define the intersection of all grounds. We say that the intersection of all grounds is the ZF-*mantle*, or simply the *mantle*, and the intersection of all ZFC-grounds is the ZFC-*mantle*. The mantle and ZFC-mantle are important objects in set-theoretic geology.

Suppose CLS, and let φ be the formula which defines all grounds uniformly as in Theorem 3.5. As mentioned in Remark 3.1, in all generic extensions of V, φ also defines all its grounds by the same way. Then we can define the *generic mantle*, which is the intersection of all grounds of all generic extensions.

Definition 7.1 (Fuchs-Hamkins-Reitz[2], Usuba[7]). Suppose CLS.

(1) The ZF-*mantle*, or simply the *mantle*, $\mathbb{M}^{\mathsf{ZF}}$ is the intersection of all grounds, and the ZFC-*mantle* $\mathbb{M}^{\mathsf{ZFC}}$ is the intersection of all ZFC-grounds.
(2) The ZF-*generic mantle*, or the *generic mantle*, $g\mathbb{M}^{\mathsf{ZF}}$ is the intersection of all grounds of all generic extensions, and the ZFC-*generic mantle* $g\mathbb{M}^{\mathsf{ZFC}}$ is the intersection of all ZFC-grounds of all generic extensions.

It is known that the (ZFC-)mantle and the (ZFC-)generic mantle are parameter free definable transitive classes (under CLS). In ZFC, it is known that the ZFC-mantle is a model of ZFC, and it coincides with the ZFC-generic mantle[2,6]. In ZF(+CLS), the generic mantle is a model of ZF, and it is a forcing invariant model[2,7].

Theorem 7.1 (Usuba[7]). *If* SVC *holds, then the generic mantle* $g\mathbb{M}^{\mathsf{ZF}}$ *is a model of* ZFC. *Moreover,* $g\mathbb{M}^{\mathsf{ZF}}$ *coincides with* $g\mathbb{M}^{\mathsf{ZFC}}$.

Let us consider the intersection of all symmetric grounds.

Definition 7.2. Suppose CLS. The *symmetric mantle* $s\mathbb{M}$ is the intersection of all symmetric grounds.

As the mantle and the generic mantle, the symmetric mantle is a parameter-free first order definable transitive class containing all ordinals. By the definitions, we have $g\mathbb{M}^{\mathsf{ZF}} \subseteq s\mathbb{M} \subseteq \mathbb{M}^{\mathsf{ZF}}$.

Proposition 7.1. *If* SVC *holds, then the symmetric mantle coincides with the generic mantle. In particular the symmetric mantle is a model of* ZFC.

Proof. Take a set x and suppose $x \notin g\mathbb{M}^{\mathsf{ZF}}$. Then there is a generic extension $V[G]$ of V and a ground W of $V[G]$ such that $x \notin W$. Note that W need not be a symmetric ground of V, but W and V are symmetric grounds of $V[G]$. By Proposition 6.1, there is M which is a common symmetric ground of W and V. Then $x \notin M$, hence $x \notin s\mathbb{M}$. □

However, beside the generic mantle, it is not known if $s\mathbb{M}$ and $\mathbb{M}^{\mathsf{ZF}}$ are always models of ZF.

8. Questions

To conclude this paper, let us pose some questions.

Question 8.1. Without any assumption, are all symmetric grounds uniformly definable?

This question is almost equivalent to the uniform definability of all grounds in ZF, this is also asked in Usuba[7].

Question 8.2. Is the symmetric mantle a model of ZF?

Question 8.3. Does the symmetric mantle always coincide with the generic mantle?

As in the context of ZFC, the downward directedness of symmetric grounds yields a positive answer of this question.

Question 8.4. Without any assumption, are all symmetric grounds downward directed?

Question 8.5. Is it consistent that the symmetric mantle is strictly smaller than the mantle?

Under $V = L$, there is no proper symmetric ground. However we do not know a choiceless model which has no proper symmetric ground.

Question 8.6. Is it consistent that V does not satisfy AC, and has no proper symmetric ground?

If there exists such a model, then SVC must fail in the model by Proposition 5.1. A similar question is:

Question 8.7. Is it consistent that V has a proper symmetric ground but has no proper ground?

If such a model exists, then AC must fail in the model, and the symmetric mantle of the model is strictly smaller than the mantle.

Acknowledgments

The author would like to thank Asaf Karagila for his many valuable comments. The author also thank the referee for many useful comments and suggestions. This research was supported by JSPS KAKENHI Grant Nos. 18K03403 and 18K03404.

References

1. A. Blass, *Injectivity, projectivity, and the axiom of choice.* Trans. Amer. Math. Soc. 255 (1979), 31–59.
2. G. Fuchs, J. D. Hamkins, J. Reitz, *Set-theoretic geology.* Ann. Pure Appl. Logic 166 (2015), no. 4, 464–501.
3. S. Grigorieff, *Intermediate submodels and generic extensions in set theory.* Annal. Math. Vol. 101 (1975), No. 3, 447–490 .
4. T. Jech, *The Axiom of Choice.* Studies in Logic and the Foundations of Mathematics, Vol. 75. North-Holland 1973.
5. A. Karagila, *The Bristol Model: an abyss called a Cohen real.* Journal of Mathematical Logic, Vol. 18 (2018), No. 02, 1850008.
6. T. Usuba, *The downward directed grounds hypothesis and very large cardinals.* J. Math. Log. 17 (2017), no. 2, 1750009.
7. T. Usuba, *Choiceless Löwenheim-Skolem property and uniform definability of grounds.* To appear in the Proceedings of the Symposium on Advances in Mathematical Logic 2018.
8. T. Usuba, *A note on Löwenheim-Skolem cardinals.* RIMS Kôkyûroku, No. 2164 (2020), 149–160.

Strong Downward Löwenheim-Skolem Theorems for stationary logics, III — Mixed support iteration

Sakaé Fuchino*, André Ottenbreit Maschio Rodrigues† and Hiroshi Sakai‡

Graduate School of System Informatics, Kobe University, Kobe, 657-8501, Japan

E-mail: *fuchino@diamond.kobe-u.ac.jp

†andreomr@gmail.com

‡hsakai@people.kobe-u.ac.jp

Continuing [Fuchino, Ottenbreit and Sakai[9,10]] and [Fuchino and Ottenbreit[11]], we further study reflection principles in connection with the Löwenheim-Skolem Theorems of stationary logics. In this paper, we mainly analyze the situations in the models obtained by mixed support iteration of a supercompact length and then collapsing another supercompact cardinal to make it $(2^{\aleph_0})^+$. We show, among other things, that the reflection down to $< 2^{\aleph_0}$ of the non-metrizability of topological spaces with small character is independent from the reflection properties studied in [Fuchino, Ottenbreit and Sakai[9,10]] and [Fuchino and Ottenbreit[11]].

Keywords: Strong Downward Löwenheim Skolem Theorem, stationary logic, generically large cardinals, mixed support iteration, Laver function, supercompact cardinal, Hamburger's Problem, continuum problem.

1. Introduction

Reflection properties of the following type are considered in various mathematical contexts:

(1.1) If a structure $\mathfrak{A}$ in the class $\mathcal{C}$ has the property $\mathcal{P}$, then there is a structure $\mathfrak{B}$ in relation $\mathcal{Q}$ to $\mathfrak{A}$ such that $\mathfrak{B}$ has the cardinality $< \kappa$ and $\mathfrak{B}$ also has the property $\mathcal{P}$.

* The first author was partially supported by JSPS Kakenhi Grant No. 20K03717. † The second author was partially supported by the Monbukagakusho (Ministry of Education, Culture, Sports, Science and Technology) Scholarship, Japan. ‡ The third author is supported by JSPS Kakenhi Grant No. 18K03397. The authors would like to thank the anonymous referee for valuable comments. An extended version of the paper with some details omitted in the published version, and with some possible further update is downloadable as: `https://fuchino.ddo.jp/papers/SDLS-III-x.pdf`

We shall call "$<\kappa$" above the *reflection point* of the reflection property (1.1). If κ is a successor cardinal μ^+, we shall also say that the reflection point of the reflection property is $\leq \mu$.

An instance of (1.1) is when $\mathcal{C}$ = "first countable topological spaces", $\mathcal{P}$ = "non-metrizable", $\mathcal{Q}$ = "subspace" and $\kappa = \aleph_2$, that is, with the reflection point $\leq \aleph_1$. In this setting, the obtained reflection statement is:

(1.2) For any first countable topological space X, if X is non-metrizable, then there is a subspace Y of X of cardinality $< \aleph_2$ such that Y is also non-metrizable.

The consistency of the statement above is still unknown. This persistently open problem about the consistency of the assertion (1.2) is called Hamburger's Problem after Peter Hamburger who asked a related question (see [Hajnal-Juhász [12]]).

The naturalness of the question can be seen in the following known partial solutions: With "first countable" replaced by "compact", the assertion (1.2) is a theorem in ZFC [Dow [6]]. With "first countable" replaced by "locally-compact", the assertion (1.2) is independent from ZFC (for the consistency we need some very large cardinal since $\square_\kappa$ for some κ implies the negation of the statement, see [Fuchino-Juhász-Szentmiklóssy-Usuba [8]]).

We shall call the following principle "Hamburger's Hypothesis" (with the reflection point $<\kappa$):

$\mathsf{HH}(<\kappa)$: For any topological space X with $\chi(x, X) < \kappa$ for all $x \in X$, if X is non-metrizable then there is a subspace Y of X of cardinality $< \kappa$ which is also non-metrizable.

Recall that the character $\chi(x, X)$ of a point x in a topological space X is the minimal possible cardinality of a neighborhood base of x in X. Without the condition on the character of points, we easily obtain a counter-example to the reflection of non-metrizability (see [Hajnal-Juhász [12]]).

Note that the original Hamburger's Problem (1.2) is equivalent to $\mathsf{HH}(<\aleph_2)$ ([Hajnal-Juhász [12]]).

The following fact will be used in the proofs of Corollary 1.8 and Proposition 4.7:

Theorem 1.1 ([Dow, Tall and Weiss [7]]). *Suppose that X is a non-metrizable space, $\delta \in \mathsf{Card}$ and $\mathbb{P} = \mathrm{Fn}(\delta, 2)$, the poset with finite conditions adding δ many Cohen reals. Then we have*

(1.3) $\Vdash_{\mathbb{P}}$ "*$\check{X}$ is non-metrizable*". □

Topological space X is considered here as a pair $X = \langle X, \tau \rangle$ where τ is the open base of the topology. Note that the family $\mathcal{O}$ of all open sets in the ground model need not to satisfy the axioms of open sets in a generic extension, while an open base remains to be an open base in the generic extension.

Let us call the posets of the form $\mathrm{Fn}(\delta, 2)$ for some ordinal δ *generalized Cohen posets.*

For a class $\mathcal{P}$ of posets, a cardinal κ is said to be *generically supercompact by* $\mathbb{P}$, if, for any $\lambda \geq \kappa$, there is a poset $\mathbb{P} \in \mathcal{P}$ such that, for a $(\mathsf{V}, \mathbb{P})$-generic $\mathbb{G}$, there are classes $j, M \subseteq \mathsf{V}[\mathbb{G}]$ such that

(1.4) M is transitive in $\mathsf{V}[\mathbb{G}]$ and $j : V \overset{\preccurlyeq}{\to} M$,

(1.5) $crit(j) = \kappa$,

(1.6) $j(\kappa) > \lambda$, and

(1.7) $j''\lambda \in M$.

Corollary 1.2. *If* κ *is generically supercompact by generalized Cohen posets, then* $\mathsf{HH}(<\kappa)$ *holds.*

Proof. Suppose that X is a non-metrizable space with

(1.8) $\chi(x, X) < \kappa$ for all $x \in X$.

Without loss of generality, $X = \langle \theta, \tau \rangle$ for some ordinal θ and an open base τ on θ. Let $\lambda \geq \theta$ be sufficiently large and let $\mathbb{P} = \mathrm{Fn}(\mu, 2)$ for some cardinal μ such that, for a $(\mathsf{V}, \mathbb{P})$-generic filer $\mathbb{G}$, there are classes j, $M \subseteq \mathsf{V}[\mathbb{G}]$ satisfying (1.4), (1.5), (1.6), and (1.7).

Let $\tau'' = \{j(O) \cap j''\theta : O \in \tau\}$. Then we have $\langle j''\theta, \tau'' \rangle$, $\langle \theta, \tau \rangle \in M$, and $M \models \langle \theta, \tau \rangle \cong \langle j''\theta, \tau'' \rangle$ by (1.7) (see, e.g. Lemma 2.5 in [Fuchino, Ottenbreit and Sakai[10]]).

By Theorem 1.1, $\mathsf{V}[\mathbb{G}] \models$ "$\langle j''\theta, \tau'' \rangle$ is non-metrizable". By (1.8), $M \models$ "$\langle j''\theta, \tau'' \rangle$ is a sub-space of $\langle j(\theta), j(\tau) \rangle$".

Thus, $M \models$ "there is a non-metrizable subspace Y of $j(X)$ of cardinality $< j(\kappa)$". By elementarity, it follows that $\mathsf{V} \models$ "there is a non-metrizable subspace Y of X of cardinality $< \kappa$". □ (Corollary 1.2)

In a model obtained as the generic extension by $\mathrm{Fn}(\kappa, 2)$ where κ is a supercompact cardinal, we have $2^{\aleph_0} = \kappa$ and κ is generically supercompact by generalized Cohen posets. Thus,

Corollary 1.3 ([Dow, Tall and Weiss[7]]). *If* ZFC + *"there is a supercompact cardinal" is consistent, then so is* $\mathsf{ZFC} + \mathsf{HH}(<2^{\aleph_0})$. □

The Strong Downward Löwenheim-Skolem Theorem $\mathsf{SDLS}^-(\mathcal{L}^{\aleph_0}_{stat}, {<}\kappa)$ for the stationary logic $\mathcal{L}^{\aleph_0}_{stat}$ down to ${<}\kappa$ is another natural reflection property. Here, the stationary logic $\mathcal{L}^{\aleph_0}_{stat}$ is a monadic second order logic whose second order variables run over countable subsets of the underlining set of the structure in question. The only second-order quantifier in the logic is '*stat*' (as well as its dual '*aa*' where the quantification "$aa\,X$" is introduced as the abbreviation of "$\neg\, stat\, X\, \neg$"). The semantics of the logic is introduced by the following step in the recursion in addition to the usual recursive definition of the semantics for first order part of the logic: for a structure $\mathfrak{A} = \langle A, ...\rangle$ and $\mathcal{L}^{\aleph_0}_{stat}$-formula $\varphi = \varphi(x_0, ..., X_0, ..., X)$ in the corresponding signature, where $X_0, ..., X$ are the second order variables in φ, as well as for $a_0, ... \in A$ and $U_0, ... \in [A]^{\aleph_0}$,

$$(1.9)\quad \begin{aligned} &\mathfrak{A} \models stat\, X\, \varphi(a_0, ..., U_0, ..., X) \\ &\Leftrightarrow\ \{U \in [A]^{\aleph_0} : \mathfrak{A} \models \varphi(a_0, ..., U_0, ..., U)\} \text{ is stationary in } [A]^{\aleph_0}. \end{aligned}$$

For a substructure $\mathfrak{B} = \langle B, ...\rangle$ of $\mathfrak{A}$, the weak variant of elementary submodel relation $\prec^-_{\mathcal{L}^{\aleph_0}_{stat}}$ between $\mathfrak{B}$ and $\mathfrak{A}$ is defined by

(1.10) $\mathfrak{B} \prec^-_{\mathcal{L}^{\aleph_0}_{stat}} \mathfrak{A} \Leftrightarrow$
$\mathfrak{B} \models \varphi(b_0, ..., b_{n-1})$ holds if and only if $\mathfrak{A} \models \varphi(b_0, ..., b_{n-1})$ holds for all $\mathcal{L}^{\aleph_0}_{stat}$-formulas $\varphi = \varphi(x_0, ..., x_{n-1})$ without free second-order variables, and for all $b_0, ..., b_{n-1} \in B$.

The reflection principle $\mathsf{SDLS}^-(\mathcal{L}^{\aleph_0}_{stat}, {<}\kappa)$ for a cardinal $\kappa \geq \aleph_2$ is defined by:

$\mathsf{SDLS}^-(\mathcal{L}^{\aleph_0}_{stat}, {<}\kappa)$: For any structure $\mathfrak{A}$ in a countable signature, there is a substructure $\mathfrak{B}$ of $\mathfrak{A}$ of cardinality ${<}\kappa$ such that $\mathfrak{B} \prec^-_{\mathcal{L}^{\aleph_0}_{stat}} \mathfrak{A}$.

In [Fuchino, Ottenbreit and Sakai[9]], we also considered the version of SDLS without '$-$' by allowing second order free variables and second order parameters in the formulas φ in (1.10). However, it is proved there that the principle $\mathsf{SDLS}(\mathcal{L}^{\aleph_0}_{stat}, {<}\kappa)$ obtained in this way for a regular κ is simply the conjunction of $\mathsf{SDLS}^-(\mathcal{L}^{\aleph_0}_{stat}, {<}\kappa)$ and $\mu^{\aleph_0} < \kappa$ for all $\mu < \kappa$.

In the standard model of PFA or under strongly Laver-generically supercompactness of a cardinal κ for proper posets (for definition of Laver-generic supercompactness, see p.112), we have the reflection principle $\mathsf{SDLS}^-(\mathcal{L}^{\aleph_0}_{stat}, {<}\aleph_2)$. Actually, $\mathsf{MA}^{+\omega_1}(\sigma\text{-closed})$ already implies this principle and strongly Laver-generically supercompactness for properness of κ

implies $\kappa = \aleph_2$ and $\mathsf{PFA}^{+\omega_1}$.

If $\mathsf{MA}^{+\omega_1}(\sigma\text{-closed})$ (or $\mathsf{PFA}^{+\omega_1}$, or $\mathsf{MM}^{+\omega_1}$, resp.) holds and $\mathbb{P}$ is $<\aleph_2$-directed closed, then we have $\Vdash_{\mathbb{P}}$ "$\mathsf{MA}^{+\omega_1}(\sigma\text{-closed})$" (or $\Vdash_{\mathbb{P}}$ "$\mathsf{PFA}^{+\omega_1}$", or $\Vdash_{\mathbb{P}}$ "$\mathsf{MM}^{+\omega_1}$" resp.) (Proposition 15 in [Fuchino and Ottenbreit[11]]).

Suppose that $\mathsf{MA}^{+\omega_1}(\sigma\text{-closed})$ holds and $2^{\aleph_0} = 2^{\aleph_1} = \aleph_2$. and there is a supercompact cardinal κ_1. Let $\mathbb{P} = \mathrm{Col}(2^{\aleph_0}, \kappa_1)$. In a generic extension by $\mathbb{P}$, we still have $\mathsf{MA}^{+\omega_1}(\sigma\text{-closed})$ by the result mentioned above, and hence also $\mathsf{SDLS}^-(\mathcal{L}^{\aleph_0}_{stat}, <\aleph_2)$. On the other hand, $\mathbb{P}$ forces κ_1 to be $(2^{\aleph_0})^+$ and makes κ_1 generically supercompact by $<\aleph_2$-closed posets (see, e.g. Lemma 4.10 in [Fuchino, Sakai and Ottenbreit[9]]). By Theorem 4.13 in [Fuchino, Sakai and Ottenbreit[9]], the assertion that $\kappa_1 = \kappa^+$ is generically supercompact by $<\kappa$-closed posets is equivalent to the Game Reflection Principle $\mathsf{GRP}^{<\kappa}(\leq \kappa)$ under $2^{<\kappa} = \kappa$. Thus, in this way, we obtain a model of a very strong reflection property with the reflection point $<2^{\aleph_0}$, together with an even stronger reflection property but with the reflection point $\leq 2^{\aleph_0}$.

$\mathsf{SDLS}^-(\mathcal{L}^{\aleph_0}_{stat}, <2^{\aleph_0})$ implies $2^{\aleph_0} = \aleph_2$ (Corollary 2.3 in [Fuchino, Sakai and Ottenbreit[10]]). This means in particular that, if the continuum should be larger than $\aleph_2$, this reflection statement is not available. In the model obtained by iterating ccc posets supercompact times with finite support along with a book-keeping provided by a Laver-function, the continuum is extremely large (e.g. in terms of existence of a saturated ideal) but the Strong Downward Löwenheim-Skolem Theorem $\mathsf{SDLS}^{int}(\mathcal{L}^{\aleph_0}_{stat}, <2^{\aleph_0})$ of the stationary logic $\mathcal{L}^{\aleph_0}_{stat}$ with internal interpretation (a weakening of $\mathsf{SDLS}^-(\mathcal{L}^{\aleph_0}_{stat}, <2^{\aleph_0})$) holds, as well as the Strong Downward Löwenheim-Skolem Theorem $\mathsf{SDLS}^{int}_+(\mathcal{L}^{\mathsf{PKL}}_{stat}, <2^{\aleph_0})$ of internal interpretation of the PKL-logic with the reflection point $\leq 2^{\aleph_0}$ (Theorem 2.10 and Proposition 3.1 in [Fuchino, Sakai and Ottenbreit[10]] for $\mathsf{SDLS}^-(\mathcal{L}^{\aleph_0}_{stat}, <2^{\aleph_0})$; Proposition 4.1 and Theorem 4.5 in [Fuchino, Sakai and Ottenbreit[10]] for $\mathsf{SDLS}^{int}_+(\mathcal{L}^{\mathsf{PKL}}_{stat}, <2^{\aleph_0})$) together with $\mathsf{MA}^{+\mu}$ for all $\mu < 2^{\aleph_0}$. The significance of $\mathsf{SDLS}^{int}_+(\mathcal{L}^{\mathsf{PKL}}_{stat}, <2^{\aleph_0})$ in this connection is that it implies that the continuum is very large (e.g. it implies that the continuum is at least weakly Mahlo).

For this model, there seems to be no way to force further to obtain a stronger reflection but with the reflection point $\leq 2^{\aleph_0}$ without destroying the reflection properties already existing in the model.

In the present paper, we show that the mixed support supercompact time iteration, roughly speaking, with Easton support mixed with the finite support, bookkept along with a Laver function together with a further collapse of the second supercompact cardinal creates a model in which "down to $< 2^{\aleph_0}$" type of reflection principles as mentioned above together with $\mathsf{GRP}^{<2^{\aleph_0}}(\leq 2^{\aleph_0})$ hold.

Modifying the finite support part of this iteration, we show the independence of $\mathsf{HH}(< 2^{\aleph_0})$ from the other strong reflection properties.

For the definition of some of the set-theoretic principles and basic facts around them remained unexplained in the present paper, the reader should consult [Fuchino, Sakai and Ottenbreit [9,10]]. These papers in extended version uploaded at the URLs given in the References may be also helpful since they contain some more details which were omitted in the submitted version of the papers.

In particular, we are going to drop the definition of $\mathsf{SDLS}_+^{int}(\mathcal{L}_{stat}^{\mathsf{PKL}}, < 2^{\aleph_0})$ and ask readers to consult [Fuchino, Sakai and Ottenbreit [10]] for details. However, we shall cite the following infinitary combinatorial characterization of this principle. This will be used in Proposition 4.7, (2) to show that this principle holds under certain instance of the two-dimensional Laver-generic large cardinal considered in Section 4.

Extending the standard notation, for sets s and t, we denote with $\mathcal{P}_s(t)$ the set

$$(1.11) \quad [\,t\,]^{|\,s\,|} = \{a \in \mathcal{P}(t) \,:\, |\,a\,| < |\,s\,|\}.$$

Lemma 1.4 (Proposition 4.1 in [Fuchino, Sakai and Ottenbreit [10]]). *For a regular cardinal $\kappa > \aleph_1$ $\mathsf{SDLS}_+^{int}(\mathcal{L}_{stat}^{\mathsf{PKL}}, < \kappa)$ is equivalent to the assertion that $(*)_{<\kappa,\lambda}^{int+\mathsf{PKL}}$ holds for all regular $\lambda \geq \kappa$ where*

$(*)_{<\kappa,\lambda}^{int+\mathsf{PKL}}$: *For any countable expansion $\mathfrak{A}$ of the structure $\langle \mathcal{H}(\lambda), \kappa, \in \rangle$ and any family $\langle S_a : a \in \mathcal{H}(\lambda)\rangle$ such that S_a is a stationary subset of $\mathcal{P}_\kappa(\mathcal{H}(\lambda))$ for all $a \in \mathcal{H}(\lambda)$, there are stationarily many $M \in \mathcal{P}_\kappa(\mathcal{H}(\lambda))$ such that $|\,\kappa \cap M\,|$ is regular, $\mathfrak{A} \restriction M \prec \mathfrak{A}$ and $S_a \cap \mathcal{P}_{\kappa\cap M}(M) \cap M$ is stationary in $\mathcal{P}_{\kappa \cap M}(M)$ for all $a \in M$.* □

We shall use freely the following "bullet notation" of names in forcing construction, introduced by Asaf Karagila[a].

If $t(x_0, ...)$ is a term in some conservative expansion of the language and the axiom system of the set theory by definitions then for a poset $\mathbb{P}$ and $\mathbb{P}$-names $\underset{\sim}{a}_0,..., t(\underset{\sim}{a}_0, ...)^\bullet$ denotes the standard $\mathbb{P}$-name $\underset{\sim}{u}$ such that

---

[a]The authors learned this extremely helpful notation in a tutorial lectures by Asaf Karagila in Kyoto at the RIMS Set Theory Workshop 2019.

(1.12) $\underset{\sim}{u}[\mathbb{G}] = t^{\mathsf{V}[\mathbb{G}]}(\underset{\sim}{a}_0[\mathbb{G}], ...)$ for any $(\mathsf{V}, \mathbb{P})$-generic filter $\mathbb{G}$
(or, more syntactically, $\Vdash_{\mathbb{P}}$ "$\underset{\sim}{u} \equiv t(\underset{\sim}{a}_0, ...)$").

For example, $\langle \underset{\sim}{a}, \underset{\sim}{b} \rangle^\bullet$ is denoted as $\mathrm{op}(\underset{\sim}{a}, \underset{\sim}{b})$ in [Kunen[17]]. $t(\underset{\sim}{a}_0, ...)$ may have infinitely many parameters. For example, if $\underset{\sim}{a}_\xi$, $\xi < \delta$ is a sequence of $\mathbb{P}$-names in the ground model, $\{\underset{\sim}{a}_\xi : \xi < \delta\}^\bullet$ may be introduced as the $\mathbb{P}$-name $\{\langle \underset{\sim}{a}_\xi, \mathbb{1}_{\mathbb{P}} \rangle : \xi < \delta\}$, while $\langle \underset{\sim}{a}_\xi : \xi < \delta \rangle^\bullet$ may be introduced as the $\mathbb{P}$-name $\{\langle \langle \check{\xi}, \underset{\sim}{a}_\xi \rangle^\bullet, \mathbb{1}_{\mathbb{P}} \rangle : \xi < \delta\}$. The choice of the exact definition of each bullet name is left to the reader. We only assume that the choice is done in a consistent way. If we want to emphasize that the bullet name $t(a_0, ...)^\bullet$ is a $\mathbb{P}$-name, we put the subscript $\mathbb{P}$ and write $t(a_0, ...)^\bullet_{\mathbb{P}}$.

For a poset $\mathbb{P}$, $\mathbb{P}$-check names of a ground model set a is represented either simply by a or with a check as $\check{a}$. If it is necessary to make clear which poset is involved, we shall also write $(a)^\surd_{\mathbb{P}}$. This representation is used, in particular, if a ground model set is given by a term. Thus we write, e.g. $(\mathcal{P}(a))^\surd_{\mathbb{P}}$, $(a \cup b)^\surd_{\mathbb{P}}$, $(\{x \in a : \varphi(x, \ldots)\})^\surd_{\mathbb{P}}$ etc.

A part of the results in the following, most of the materials in Section 5 in particular, have been presented in the PhD thesis [Ottenbreit Maschio Rodrigues[18]] of the second author, although some arguments and details are treated differently from those in the PhD thesis.

2. Reflection number of Hamburger's Hypothesis

As usual, $\mathfrak{b}$ denotes the bounding number which is defined as the minimal possible cardinality of a subset of ${}^\omega\omega$ which is unbounded with respect to $\leq^*$ (coordinate-wise comparison modulo finite).

Topological spaces constructed in Remark 12.6 in [van Douwen[5]] witness the following theorem:

Theorem 2.1 ([van Douwen[5]]). *There is a Moore space X of cardinality $\mathfrak{b}$ such that X is not collectionwise Hausdorff (and hence non-metrizable) but all subspaces of X of cardinality $< \mathfrak{b}$ are metrizable.* □

The proof of Theorem 2.1 gives a construction of topological spaces $X_{\mathcal{F}}$ for $\mathcal{F} \subseteq {}^\omega\omega$. The construction is enough absolute and these spaces satisfy the properties that $X_{\mathcal{F}}$ is a developable normal Hausdorff space, $X_{\mathcal{F}}$ is a subspace of $X_{\mathcal{F}'}$ for $\mathcal{F} \subseteq \mathcal{F}' \subseteq {}^\omega\omega$, and $X_{\mathcal{F}}$ is metrizable if and only if $\mathcal{F}$ is bounded in ${}^\omega\omega$ (with respect to $\leq^*$).

Corollary 2.2. *There is a non-metrizable Moore space $X = \langle X, \tau\rangle$ such that $\Vdash_{\mathbb{P}}$ "$\check{X}$ is metrizable" for a σ-centered poset $\mathbb{P}$.*

Proof. Let $X = X_{\mathcal{F}}$ for an unbounded family $\subseteq {}^{\omega}\omega$. Let $\mathbb{P}$ be the Hechler forcing then $\Vdash_{\mathbb{P}}$ "$\check{\mathcal{F}}$ is bounded". Thus, $\Vdash_{\mathbb{P}}$ "$X_{\check{\mathcal{F}}}$ is metrizable". By the absoluteness of the definition of $X_{\mathcal{F}}$, we have $\Vdash_{\mathbb{P}}$ "$\check{X} = X_{\check{\mathcal{F}}}$". ❑ (Corollary 2.2)

The reflection number $\mathfrak{Refl}_{\mathsf{HH}}$ of Hamburger's Hypothesis is defined by:

$$(2.1) \qquad \mathfrak{Refl}_{\mathsf{HH}} = \begin{cases} \text{the minimal cardinal } \kappa \text{ such that,} & \\ \text{for any first countable non-metrizable} & \\ \text{topological space } X, \text{ there is a non-} & \\ \text{metrizable subspace } Y \text{ of } X \text{ of} & \\ \text{cardinality} < \kappa; & \text{if such } \kappa \text{ exists,} \\ \infty; & \text{otherwise.} \end{cases}$$

Lemma 2.3. (1) $\mathfrak{b} < \mathfrak{Refl}_{\mathsf{HH}} \leq \infty$.

(2) $\mathfrak{Refl}_{\mathsf{HH}} = \infty$ *is consistent.*

(3) ([**Bagaria and Magidor**[1]]) $\mathfrak{Refl}_{\mathsf{HH}} \leq$ *the least ω_1-strongly compact cardinal (if it exists).*

Proof. (1): By Theorem 2.1.

(2): This holds if $\Box_\kappa$ holds for cofinally many κ (in Card) — actually $\mathsf{ADS}^-(\kappa)$ for class many regular uncountable κ is enough (see Proposition 6.3 in [Fuchino, Juhász et al.[8]])).

(3): See: [Bagaria and Magidor[1]]. ❑ (Lemma 2.3)

3. Preservation and non-preservation of stationarity of subsets of $\mathcal{P}_\kappa(\lambda)$

In the following, we show that the closedness of posets cannot be used to establish reflection principles concerning the stationarity of subsets of $\mathcal{P}_\kappa(\lambda)$ for $\kappa > \aleph_1$ in the generic extensions. At least, not in a straightforward generalization of the usage of σ-closed posets in a forcing argument to obtain reflection properties on stationarity of subsets of $\mathcal{P}_{\aleph_1}(\lambda)$ in the generic extensions.

Actually, the examples of preservation and non-preservation of stationarity of subsets of $\mathcal{P}_\kappa(\lambda)$ in this section explain, why we need a mixed support iteration plus one further step with chain condition in connection with the following Lemma 3.1 to establish (some of the) results in Section 6 but not in a much simpler way.

It is well-known that ccc posets and σ-closed posets are proper. This means that such posets preserve stationarity of subsets of $\mathcal{P}_{\aleph_1}(\lambda)$ for any uncountable λ. For posets with κ-cc for regular cardinal $\kappa > \aleph_1$ we still have a corresponding lemma:

Lemma 3.1. *Suppose that κ is a regular uncountable cardinal and $\lambda \geq \kappa$. If $S \subseteq \mathcal{P}_\kappa(\lambda)$ is stationary and $\mathbb{P}$ is a κ-cc poset, then we have $\Vdash_{\mathbb{P}}$ "$\check{S}$ is stationary".*

The standard proof for the case $\kappa = \aleph_1$ also works for this general lemma.

In contrast, κ-closed poset can destroy stationarity of ground model stationary set $\subseteq \mathcal{P}_\kappa(\lambda)$ if $\kappa > \aleph_1$. This makes consistency proofs of stationary reflection of stationary subsets of $\mathcal{P}_\kappa(\lambda)$ for $\kappa \geq \aleph_2$ more involved. In the following, we shall examine situations where the stationarity of some subset of $\mathcal{P}_\kappa(\lambda)$ for $\kappa \geq \aleph_2$ is not preserved by a standard κ-closed poset.

For cardinal κ and a regular cardinal $\nu < \kappa$ we denote

$$E^\kappa_\nu = \{\alpha \in \kappa : cf(\alpha) = \nu\}. \tag{3.1}$$

The following Lemma is used for our first example of non-preservation of stationarity in Proposition 3.3.

Lemma 3.2. *Suppose that κ is a regular cardinal with $\kappa \geq \aleph_2$ and $X \supseteq \kappa^+$. Then, for any distinct regular ν, $\mu < \kappa$,*

$$S = \{x \in \mathcal{P}_\kappa(X) : \kappa \cap x \in E^\kappa_\nu,\ cf(\sup(\kappa^+ \cap x)) = \mu\} \tag{3.2}$$

is stationary in $\mathcal{P}_\kappa(X)$.

Proof. Suppose that $C \subseteq \mathcal{P}_\kappa(X)$ is a club. Let $f : [X]^{<\aleph_0} \to X$ be such that $C\ell^*(f) = \{x \in \mathcal{P}_\kappa(X) : x \cap \kappa \in \kappa,\ x \text{ is closed with respect to } f\} \subseteq C$.

Let $X_0 \subseteq X$ be such that X_0 is closed with respect to f and $\kappa^+ \cap X_0 \in E^{\kappa^+}_\mu$. Let $\delta = \kappa^+ \cap X_0$.

Let $\langle x_\xi : \xi < \nu \rangle$ be a continuously increasing sequence in $\mathcal{P}_\kappa(X_0)$ such that

(3.3) x_ξ is closed with respect to f for all $\xi < \nu$;
(3.4) $\sup(\kappa^+ \cap x_0) = \delta$; and
(3.5) $\sup(\kappa \cap x_\xi) + 1 \subseteq x_{\xi+1}$ for all $\xi < \nu$.

Note that this construction is possible since κ is regular and ν, $\mu < \kappa$.

Let $x = \bigcup_{\xi<\nu} x_\xi$. Then

(3.6) x is closed with respect to f; (by (3.3))

(3.7) $\sup(\kappa^+ \cap x) = \delta$; (by (3.4))

(3.8) $\kappa \cap x \in \kappa$ and $cf(\kappa \cap x) = \nu$. (by (3.5))

$x \in S$ by (3.7) and (3.8). $x \in C\ell^*(f)$ by (3.6) and (3.8). Thus we have $\emptyset \neq S \cap C\ell^*(f) \subseteq S \cap C$. □ (Lemma 3.2)

For a regular cardinal, $\mathrm{Add}(\kappa)$ denotes the set ${}^{\kappa>}2$ with the reverse inclusion. We denote with $\mathrm{Col}(\kappa, \kappa^+)$ the set ${}^{\kappa>}\kappa^+$ with the reverse inclusion. $\mathrm{Add}(\kappa)$ and $\mathrm{Col}(\kappa, \kappa^+)$ are forcing equivalent to $\mathrm{Fn}(\kappa, 2, \kappa)$ and $\mathrm{Fn}(\kappa, \kappa^+, \kappa)$ in Kunen's notation in [Kunen[17]], respectively. The posets isomorphic to latter two posets are also denoted as $\mathrm{Col}(\kappa, \{\kappa\})$ and $\mathrm{Col}(\kappa, \{\kappa^+\})$ respectively, in the notation of [Kanamori[14]]. Both of the posets are κ-closed. $\mathrm{Add}(\kappa)$ adds a new subset of κ while κ^+ is preserved if $\kappa^{<\kappa} = \kappa$. $\mathrm{Col}(\kappa, \kappa^+)$ collapses κ^+ and makes it of cardinality and cofinality κ.

Proposition 3.3. *Suppose that κ is a regular cardinal $\geq \aleph_2$ and $X \supseteq \kappa^+$. Then, there is a stationary $S \subseteq \mathcal{P}_\kappa(X)$, such that $\Vdash_{\mathrm{Col}(\kappa,\kappa^+)}$ "$\check{S}$ is not stationary in $\mathcal{P}_\kappa(X)$".*

Proof. In V, let

(3.9) $S = \{x \in \mathcal{P}_\kappa(X) : x \cap \kappa$ and $\sup(x \cap \kappa^+)$ are limit ordinals, and $cf(x \cap \kappa) \neq cf(\sup(x \cap \kappa^+))\}$.

S is a stationary subset of $\mathcal{P}_\kappa(X)$ by Lemma 3.2. We show that $\mathrm{Col}(\kappa, \kappa^+)$ forces that S is not stationary.

Suppose that $\mathbb{G}$ is a $(\mathsf{V}, \mathrm{Col}(\kappa, \kappa^+))$-generic filter. Note that, by $<\kappa$-closedness, $\mathrm{Col}(\kappa, \kappa^+)$ does not add any new sets of size $<\kappa$. Thus $\mathcal{P}_\kappa(X)^{\mathsf{V}} = \mathcal{P}_\kappa(X)^{\mathsf{V}[\mathbb{G}]}$, all cofinalities $<\kappa$ are preserved in the generic extension $\mathsf{V}[\mathbb{G}]$, and $cf(\mu) = \kappa$ in $\mathsf{V}[\mathbb{G}]$ for $\mu = (\kappa^+)^{\mathsf{V}}$.

In $\mathsf{V}[\mathbb{G}]$, let $\langle \gamma_\alpha : \alpha < \kappa \rangle$ be a continuously increasing sequence of ordinals cofinal in the ordinal μ. Let

(3.10) $C = \{x \in \mathcal{P}_\kappa(X) : x \cap \kappa$ and $\sup(x \cap \mu)$ are limit ordinals, and $\sup(x \cap \mu) = \gamma_{x \cap \kappa}\}$.

Then C is a club in $\mathcal{P}_\kappa(X)$ and $C \cap S = \emptyset$. □ (Proposition 3.3)

In Proposition 3.3, the crucial fact which made the set S non-stationary in the generic extension was that the cardinal κ^+ is collapsed to be an ordinal of cofinality κ. However, stationarity of $\mathcal{P}_\kappa(\lambda)$ can be also destroyed by a $<\kappa$-closed forcing without collapsing cardinals:

Proposition 3.4. *Suppose that κ is a supercompact and $|X| \geq 2^\kappa$. Then there is a stationary $S \subseteq \mathcal{P}_\kappa(X)$ such that $\Vdash_{\mathrm{Add}(\kappa)}$ "$\check{S}$ is not a stationary subset of $\mathcal{P}_\kappa(X)$".*

Note that $|\mathrm{Add}(\kappa)| = \kappa$ since κ is inaccessible and hence $\mathrm{Add}(\kappa)$ is κ^+-cc. Thus $\mathrm{Add}(\kappa)$ here preserves cardinals and cofinality.

Proof. Let $\lambda = |X|$. Without loss of generality, we may assume that $X = \lambda$. In V, let $\vec{B} = \langle B_\alpha : \alpha < \lambda\rangle$ be an enumeration of $\mathcal{P}(\kappa)$ and let

$$(3.11)\quad S = \{x \in \mathcal{P}_\kappa(\lambda) : \text{(a) } \kappa \cap x \in \kappa, \text{ and } \text{(b) } \{B_\alpha \cap (x \cap \kappa) : \alpha \in x\} = \mathcal{P}(\kappa \cap x)\}.$$

Claim 3.4.1. *$S \in \mathcal{U}$ for any normal ultrafilter $\mathcal{U}$ over $\mathcal{P}_\kappa(\lambda)$.*

$\vdash$ Suppose that $\mathcal{U}$ is a normal ultrafilter over $\mathcal{P}_\kappa(\lambda)$. It is enough to show that $j_{\mathcal{U}}{}''\lambda \in j_{\mathcal{U}}(S)$ where $j_{\mathcal{U}} : \mathsf{V} \overset{\preceq}{\to} M$ is the elementary embedding induced by $\mathcal{U}$.

We have

$$(3.12)\quad (j_{\mathcal{U}}{}''\lambda) \cap j_{\mathcal{U}}(\kappa) = \{j_{\mathcal{U}}(\alpha) : \alpha < \lambda,\ j_{\mathcal{U}}(\alpha) < j_{\mathcal{U}}(\kappa)\} = \{j_{\mathcal{U}}(\alpha) : \alpha < \kappa\} = \{\alpha : \alpha \in \kappa\} = \kappa \in j_{\mathcal{U}}(\kappa).$$

For $\beta \in j_{\mathcal{U}}{}''\lambda$ with $\beta = j_{\mathcal{U}}(\alpha)$ for $\alpha \in \lambda$, $j_{\mathcal{U}}(\vec{B})(\beta) \cap (j_{\mathcal{U}}{}''\lambda \cap j_{\mathcal{U}}(\kappa)) = j_{\mathcal{U}}(B_\alpha) \cap \kappa = B_\alpha$.

Thus we have

$$(3.13)\quad \{j_{\mathcal{U}}(\vec{B})(\beta) \cap (j_{\mathcal{U}}{}''\lambda \cap j_{\mathcal{U}}(\kappa)) : \beta \in j''\lambda\} = \{B_\alpha : \alpha < \lambda\} = \mathcal{P}(\kappa) = \mathcal{P}(j_{\mathcal{U}}{}''\lambda \cap j(\kappa)).$$

By elementarity, (3.12) and (3.13) imply $j_{\mathcal{U}}{}''\lambda \in j(S)$. $\dashv$ (Claim 3.4.1)

Since any normal filter over $\mathcal{P}_\kappa(\lambda)$ contains all club sets and hence it consists of stationary sets, and since there *are* normal ultrafilters over $\mathcal{P}_\kappa(\lambda)$ because κ is supercompact, we conclude that S is a stationary subset of $\mathcal{P}_\kappa(\lambda)$.

Thus the next Claim shows that S is as desired:

Claim 3.4.2. $\Vdash_{\mathrm{Add}(\kappa)}$ *"S is not stationary in $\mathcal{P}_\kappa(\lambda)$".*

$\vdash$ Suppose that $\mathbb{G}$ is a $(\mathsf{V}, \mathrm{Add}(\kappa))$-generic filter. In $\mathsf{V}[\mathbb{G}]$, we have $\bigcup\mathbb{G} : \kappa \to 2$. Let $A = (\bigcup\mathbb{G})^{-1}{}''\{1\}$. By genericity, A is a new subset of κ. Let $F : \lambda \to \kappa$ be defined by $F(\alpha) = \min(B_\alpha \mathbin{\triangle} A)$ for $\alpha < \lambda$. F is well-defined since $B_\alpha \neq A$ for all $\alpha < \lambda$. We show that $S \cap C_F = \emptyset$ where C_F is the club set defined by

(3.14) $\quad C_F = \{x \in \mathcal{P}_\kappa(\lambda) : x \text{ is closed with respect to } F\}.$

Suppose that $x \in S$. By (b) in (3.11) and since $A \cap x \in \mathcal{P}(x \cap \kappa)^{\mathsf{V}}$, there is an $\alpha^* \in x$ such that $B_{\alpha^*} \cap x = A \cap x$. But this implies that $F(\alpha^*) = min(B_\alpha \triangle A) \notin x$. Thus, x is not closed with respect to F and $x \notin C_F$. $\dashv$ (Claim 3.4.2)

$\square$ (Proposition 3.4)

The non-preservation of stationarity of subsets of $\mathcal{P}_\kappa(\lambda)$ along the line of the results above is further studied in [Sakai [19]].

4. Two dimensional Laver-generic large cardinals

For properties $\mathfrak{P}$ and $\mathfrak{Q}$ of posets, a cardinal κ is *Laver-generically supercompact for* $(\mathfrak{P}, \mathfrak{Q})$ if, for any poset $\mathbb{P}$ with $\mathbb{P} \models \mathfrak{P}$ and $(\mathsf{V}, \mathbb{P})$-generic $\mathbb{G}$, there are a $\mathbb{P}$-name $\underset{\sim}{\mathbb{Q}}$ of a poset with $\Vdash_{\mathbb{P}}$ "$\underset{\sim}{\mathbb{Q}} \models \mathfrak{Q}$" and a $(\mathsf{V}, \mathbb{P} * \underset{\sim}{\mathbb{Q}})$-generic $\mathbb{H}$ with $i''\mathbb{G} \subseteq \mathbb{H}$ where $i : \mathbb{P} \to \mathbb{P} * \underset{\sim}{\mathbb{Q}}$ is the canonical complete embedding, such that there are $j, M \subseteq \mathsf{V}[\mathbb{H}]$ where:

(4.1) $\quad M$ is a transitive class in $\mathsf{V}[\mathbb{H}]$;

(4.2) $\quad j : \mathsf{V} \overset{\preceq}{\to} M$;

(4.3) $\quad crit(j) = \kappa$ and $j(\kappa) > \lambda$;

(4.4) $\quad \mathbb{P}, \mathbb{H} \in M$, and

(4.5) $\quad j''\lambda \in M$.

κ is *strongly Laver-generically supercompact for* $(\mathfrak{P}, \mathfrak{Q})$ if M in the definition of the Laver-generic supercompactness for $(\mathfrak{P}, \mathfrak{Q})$ additionally satisfies

(4.6) $\quad ([M]^{\aleph_0})^{\mathsf{V}[\mathbb{H}]} \subseteq M.$

If $\mathfrak{P}$ solely consists of trivial forcing, we shall say "*strongly Laver-generically supercompact for* $\mathfrak{Q}$" instead of "strongly Laver-generically supercompact for $(\mathfrak{P}, \mathfrak{Q})$".

In the following, both of the properties $\mathfrak{P}$ and $\mathfrak{Q}$ considered in connection with the Laver-generic supercompactness imply the properness of the poset. In such a case, the model of the Laver-generic supercompactness constructed by forcing starting from a supercompact cardinal usually satisfies this strong version of Laver-generic supercompactness as well. This is because of the following well-known fact:

Lemma 4.1. *Suppose that $M \subseteq \mathsf{V}$ is an inner model with*

(4.7) $\quad [M]^{\aleph_0} \subseteq M.$

If $\mathbb{P} \in M$ is proper and $\mathbb{G}$ is a $(\mathsf{V}, \mathbb{P})$-generic filter, then we have

(4.8) $\quad ([M[\mathbb{G}]]^{\aleph_0})^{\mathsf{V}[\mathbb{G}]} \subseteq M[\mathbb{G}].$

Proof. Let $a \in ([M[\mathbb{G}]]^{\aleph_0})^{\mathsf{V}[\mathbb{G}]}$ and $\underset{\sim}{a}$ be a $\mathbb{P}$-name of a. In V, let θ be a sufficiently large regular cardinal and let $N \prec \mathcal{H}(\theta)$ and $\mathbb{p} \in \mathbb{G}$ be such that

(4.9) $\quad |N| = \aleph_0;$

(4.10) $\quad \underset{\sim}{a}, \mathbb{P} \in N;$

(4.11) For any maximal antichain $A \subseteq \mathbb{P}$ with $A \in N$, $A \cap N$ is predense below $\mathbb{p}$.

Let $\underset{\sim}{f} \in N$ be a $\mathbb{P}$-name such that:
$\Vdash_{\mathbb{P}}$ "$\underset{\sim}{f} : \omega \to \underset{\sim}{a}$ is a surjection". For each $n \in \omega$, let $A_n \in M$ be a maximal antichain such that, for each $\mathbb{r} \in A_n$, there is a $\mathbb{P}$-name $\underset{\sim}{b}_{n,\mathbb{r}} \in M$ such that $\mathbb{r} \Vdash_{\mathbb{P}}$ "$\underset{\sim}{f}(n) \equiv \underset{\sim}{b}_{n,\mathbb{r}}$". Note that we have $\underset{\sim}{b}_{n,\mathbb{r}} \in M \cap N$ if $\mathbb{r} \in N$, since $\underset{\sim}{b}_{n,\mathbb{r}}$ is uniquely determined for each $\mathbb{r} \in A_n$.

Let $\underset{\sim}{a}^* = \{\langle \underset{\sim}{b}_{n,\mathbb{r}}, \mathbb{r}\rangle : n \in \omega, \mathbb{r} \in A_n \cap N\}$. Then $\underset{\sim}{a}^* \in M$ by (4.7) and $\underset{\sim}{a}^*[\mathbb{G}] = \underset{\sim}{a}[\mathbb{G}]$. Thus, $a = \underset{\sim}{a}[\mathbb{G}] \in M[\mathbb{G}]$. □ (Lemma 4.1)

The condition (4.6) can be even replaced with

(4.6′) $\quad ([M]^{j(\kappa)})^{\mathsf{V}[\mathbb{H}]} \subseteq M,$

if we consider Laver-generic superhugeness instead of Laver-generic supercompactness.

This can be seen by means of the following:

Lemma 4.2. *Suppose that M is an inner model of V with*

(4.12) $\quad V \models$ "$[M]^{\mu} \subseteq M$"

for a regular μ. If $\mathbb{P} \in M$ is μ^+-cc, then, for any $(\mathsf{V}, \mathbb{P})$-generic $\mathbb{G}$, we have

(4.13) $\quad ([M[\mathbb{G}]]^{\mu})^{\mathsf{V}[\mathbb{G}]} \subseteq M[\mathbb{G}].$

Proof. Suppose $g \in ({}^{\mu}M[\mathbb{G}])^{\mathsf{V}[\mathbb{G}]}$. We show that $g \in M[\mathbb{G}]$. Let $\underset{\sim}{g}$ be a $\mathbb{P}$-name of g. For each $\xi < \mu$, there is a maximal pairwise incompatible

$A_\xi \subseteq \mathbb{P}$ such that, for each $\mathbb{p} \in A_\xi$, there is a $\mathbb{P}$-name $\underset{\sim}{a}_{\xi,\mathbb{p}} \in M$ such that $\mathbb{p} \Vdash_{\mathbb{P}}$ "$\underset{\sim}{g}(\xi) \equiv \underset{\sim}{a}_{\xi,\mathbb{p}}$". By the μ^+-cc of $\mathbb{P}$, we have $|\,A_\xi\,| \leq \mu$ and hence $A_\xi \in M$ by (4.12).

Let

$$(4.14) \quad \underset{\sim}{a}_\xi = \{\langle \underset{\sim}{b}, \mathbb{q}\rangle \,:\, \mathbb{q} \leq_{\mathbb{P}} \mathbb{p} \text{ for some } \mathbb{p} \in A_\xi,\ \underset{\sim}{b} \text{ is a canonical } \mathbb{P}\text{-name with } \mathbb{q} \Vdash_{\mathbb{P}} \text{“}\underset{\sim}{b} \in \underset{\sim}{a}_{\xi,\mathbb{p}}\text{”}\}.$$

$\underset{\sim}{a}_\xi \in M$ since it is definable from $\leq \mu$ many parameters from M and by (4.12). It is also clear by the definition above that $\Vdash_{\mathbb{P}}$ "$\underset{\sim}{g}(\xi) \equiv \underset{\sim}{a}_\xi$". Let

$$(4.15) \quad \underset{\sim}{g}^* = \{\langle \langle \check{\xi}, \underset{\sim}{a}_\xi\rangle^\bullet_{\mathbb{P}}, \mathbb{1}_{\mathbb{P}}\rangle \,:\, \xi < \mu\}.$$

Then $\underset{\sim}{g}^* \in M$ by (4.12) and $M[\mathbb{G}] \ni \underset{\sim}{g}^*[\mathbb{G}] = \underset{\sim}{g}[\mathbb{G}] = g$. □ (Lemma 4.2)

Lemma 4.2 is used with a generic superhuge cardinal to produce the strong generic superhugeness. This can be seen as follows:

Suppose that κ is a superhuge cardinal and $\vec{\mathbb{P}} = \langle \mathbb{P}_\alpha, \underset{\sim}{\mathbb{Q}}_\beta \,:\, \alpha \leq \kappa, \beta < \kappa\rangle$ be an iteration with

$$(4.16) \quad |\,\mathbb{P}_\alpha\,| < \kappa \text{ for all } \alpha < \kappa.$$

Let $\mathbb{G}_\kappa$ be a $(\mathsf{V}, \mathbb{P}_\kappa)$-generic filter.

For a given cardinal λ, let $j : \mathsf{V} \overset{\preccurlyeq}{\to} M$ be an elementary embedding into an inner model M of V such that $crit(j) = \kappa$, $j(\kappa) > \lambda$ and $[M]^{j(\kappa)} \subseteq M$. Let $\mathbb{P}^* = j(\mathbb{P}_\kappa)$. By elementarity $\mathbb{P}^*$ is the $j(\kappa)$-th iterand of the iteration $j(\vec{\mathbb{P}})$ in M. There is the canonical complete embedding $i : \mathbb{P}_\kappa \overset{\lessdot}{\to} \mathbb{P}^*$. Let $\mathbb{G}^*$ be $(\mathsf{V}, \mathbb{P}^*)$-generic filter with

$$(4.17) \quad i''\mathbb{G}_\kappa \subseteq \mathbb{G}^*.$$

By (4.16) and elementarity, we have $M \models$ "$|\,\mathbb{P}^*\,| \leq j(\kappa)$". Hence $|\,\mathbb{P}^*\,| \leq j(\kappa)$ and thus $\mathbb{P}^*$ has $j(\kappa)^+$-cc. By (4.17), j can be lifted to

$$(4.18) \quad \tilde{j} : V[\mathbb{G}_\kappa] \overset{\preccurlyeq}{\to} M[\mathbb{G}^*];\ \underset{\sim}{a}[\mathbb{G}_\kappa] \mapsto j(\underset{\sim}{a})[\mathbb{G}^*]$$

for $\mathbb{P}_\kappa$-names $\underset{\sim}{a}$. Now by Lemma 4.2, we have $([M[\mathbb{G}^*]]^{j(\kappa)})^{\mathsf{V}[\mathbb{G}^*]} \subseteq M[\mathbb{G}^*]$.

Based on the observations above, we define κ to be *strongly Laver-generically superhuge for the pair of properties* $(\mathfrak{P}, \mathfrak{Q})$ if, for any poset $\mathbb{P}$ with $\mathbb{P} \models \mathfrak{P}$ and $(\mathsf{V}, \mathbb{P})$-generic $\mathbb{G}$, there are a $\mathbb{P}$-name $\underset{\sim}{\mathbb{Q}}$ of a poset with

$\Vdash_{\mathbb{P}}$ "$\underset{\sim}{\mathbb{Q}} \models \mathfrak{Q}$" and a $(\mathsf{V}, \mathbb{P}*\underset{\sim}{\mathbb{Q}})$-generic $\mathbb{H}$ with $i''\mathbb{G} \subseteq \mathbb{H}$ where $i : \mathbb{P} \to \mathbb{P}*\underset{\sim}{\mathbb{Q}}$ is the canonical complete embedding, such that there are $j, M \subseteq \mathsf{V}[\mathbb{H}]$ with (4.1) $\sim$ (4.4) and (4.6′).

If $\mathfrak{P}$ solely consists of trivial forcing, we shall say "*strongly superhuge for* $\mathfrak{Q}$" instead of "strongly Laver-generically superhuge for $(\mathfrak{P}, \mathfrak{Q})$".

The following is trivial.

Lemma 4.3. *Suppose that* $\mathbb{P} \models \mathfrak{P}_0$ *implies* $\mathbb{P} \models \mathfrak{P}_1$ *and* $\mathbb{P} \models \mathfrak{Q}_1$ *implies* $\mathbb{P} \models \mathfrak{Q}_0$ *for all posets* $\mathbb{P}$ *(i.e. these implications are theorems in* ZFC*). If* κ *is (strongly) Laver-generically supercompact/superhuge for* $(\mathfrak{P}_1, \mathfrak{Q}_1)$*, then* κ *is (strongly) Laver-generically supercompact/huge for* $(\mathfrak{P}_0, \mathfrak{Q}_0)$. □

We call a property $\mathfrak{P}$ of posets *iterable* if we can prove in ZFC that

(4.19) $\mathbb{P} * \underset{\sim}{\mathbb{Q}} \models \mathfrak{P}$ for any poset $\mathbb{P}$ with $\mathbb{P} \models \mathfrak{P}$ and $\mathbb{P}$-name $\underset{\sim}{\mathbb{Q}}$ of a poset with $\Vdash_{\mathbb{P}}$ "$\underset{\sim}{\mathbb{Q}} \models \mathfrak{P}$".

Proposition 4.4. *Suppose that* $\mathfrak{P}$ *is the property "forcing equivalent to a poset of the form* $\mathrm{Col}(\kappa, \mu)$ *for some* μ*"*[h]*,* $\mathfrak{Q}$ *is iterable and we can prove (in* ZFC*) that*

(4.20) $\forall P\,(P$ *is a* σ*-directed closed poset* $\to P \models \mathfrak{Q})$.

If κ *is strongly Laver-generically supercompact for* $(\mathfrak{P}, \mathfrak{Q})$*, then, for any* $\mu \geq \kappa$*, and for* $\mathbb{P} = \mathrm{Col}(\kappa, \mu)$*, we have*

(4.21) $\Vdash_{\mathbb{P}}$ "*κ is generically supercompact by posets satisfying* $\mathfrak{Q}$ ".

Proof. The proof is a typical application of the master condition argument.

Note that $\mathbb{P} \models \mathfrak{P}$. Let $\mathbb{G}$ be an arbitrary $(\mathsf{V}, \mathbb{P})$-generic filter. We have to show that $\mathsf{V}[\mathbb{G}] \models$ "κ is generic supercompact by posets with $\mathfrak{Q}$".

Let $\lambda \geq \theta$, $\mu^{<\kappa}$ and let $\underset{\sim}{\mathbb{Q}}$ be a $\mathbb{P}$-name with $\Vdash_{\mathbb{P}}$ "$\underset{\sim}{\mathbb{Q}}$ is a poset with $\mathfrak{Q}$" such that there is a $(\mathsf{V}, \mathbb{P} * \underset{\sim}{\mathbb{Q}})$-generic filter $\mathbb{H}$ such that $i''\mathbb{G} \subseteq \mathbb{H}$ for the canonical complete embedding $i : \mathbb{P} \to \mathbb{P} * \underset{\sim}{\mathbb{Q}}$, with j, $M \subseteq \mathsf{V}[\mathbb{H}]$ such that (4.1) $\sim$ (4.5) and (4.6) hold.

By (4.4), we have $\mathbb{G} \in M$. Let $\mathbb{R} = j(\mathbb{P})$. By elementarity,

(4.22) $M \models$ "$\mathbb{R}$ is $<j(\kappa)$-directed closed".

---

[h]In this section, we are back to Kanamori's notation of collapsing posets. $\mathrm{Col}(\kappa, \lambda)$ for an inaccessible λ is thus the poset collapsing all cardinals strictly between κ and λ by conditions of size $<\kappa$.

By (4.22) and (4.6), $\mathsf{V} \models$ "$\mathbb{R}$ is σ-directed closed". Thus, $\mathsf{V} \models$ "$\mathbb{R} \models \mathfrak{Q}$" by (4.21).

By (4.3), $M \models |j''\mathbb{G}| < j(\kappa)$. Hence, there is (a master condition) $\mathbb{r} \in \mathbb{R}$ such that $M \models \mathbb{r} \leq_{\mathbb{R}} j''\mathbb{G}$ by (4.22). Let $\mathbb{K}$ be a $(\mathsf{V}[\mathbb{H}], \mathbb{R})$-generic filter with $\mathbb{r} \in \mathbb{K}$. Then

$$(4.23) \quad \tilde{j} : \mathsf{V}[\mathbb{G}] \overset{\prec}{\to} M[\mathbb{K}];\ \underset{\sim}{a}^{\mathbb{G}} \mapsto j(\underset{\sim}{a})^{\mathbb{K}}$$

is well defined and $j \subseteq \tilde{j}$. In particular we have $\kappa = crit(\tilde{j})$, $\tilde{j} > \lambda$ and $\tilde{j}''\lambda \in M[\mathbb{K}]$.

By (4.22) and (4.5), we have $\mathsf{V}[\mathbb{H}] \models$ "$\mathbb{R}$ is σ-directed closed". Thus, in $\mathsf{V}[\mathbb{G}]$, letting $\mathbb{Q} = \underset{\sim}{\mathbb{Q}}[\mathbb{G}]$, with the $\mathbb{Q}$-name $\underset{\sim}{\mathbb{R}}$ corresponding to $\mathbb{R}$ such that $\Vdash_{\mathbb{Q}}$ "$\underset{\sim}{\mathbb{R}}$ is $<\theta$-directed closed", $\mathbb{Q} * \underset{\sim}{\mathbb{R}}$ satisfies $\mathfrak{Q}$ and it induces a generic elementary embedding for generic λ-supercompactness. Since $\mathbb{G}$ was arbitrary, it follows that (4.21) holds. □ (Proposition 4.4)

Corollary 4.5. *Suppose that κ is strongly Laver-generically supercompact for $(\mathfrak{P}, \mathfrak{Q})$ where $\mathfrak{P}$ is the property "forcing equivalent to a poset of the form* $\mathrm{Col}(\kappa, \mu)$ *for some μ" and $\mathfrak{Q}$ is "proper". Suppose further that $\kappa_1 > \kappa$ is a supercompact and let $\mathbb{P} = \mathrm{Col}(\kappa, \kappa_1)$. Then*

(a) $\Vdash_{\mathbb{P}}$ "*κ is generically supercompact by proper posets*";

(b) $\Vdash_{\mathbb{P}}$ "*κ^+ is generically supercompact by $<\kappa$-closed posets*";

(c) $\Vdash_{\mathbb{P}}$ "$\mathsf{SDLS}^{int}_{+}(\mathcal{L}^{\aleph_0}_{stat}, <\kappa)$"; *and*

(d) $\Vdash_{\mathbb{P}}$ "$\mathsf{GRP}^{<\kappa}(\leq\kappa)$".

Proof. (a): By Proposition 4.4.
(b): By Lemma 4.10 in [Fuchino, Sakai and Ottenbreit[9]].
(c): By (a) and, Theorem 2.10 and Propositions 3.1 in [Fuchino, Sakai and Ottenbreit[10]].
(d): By (b) above and Lemma 4.11 in [Fuchino, Sakai and Ottenbreit[9]].
□ (Corollary 4.5)

Proposition 4.6. *Suppose that $\mathfrak{P}$ is the property "forcing equivalent to a poset of the form* $\mathrm{Col}(\kappa, \mu)$ *for some μ" and $\mathfrak{Q}$ the property of posets such that*

$$(4.24) \quad \forall P \forall \underset{\sim}{Q}\, (P \models \mathfrak{P} \wedge \Vdash_P \text{"}\underset{\sim}{Q} \text{ is a } <\theta\text{-directed closed poset"} \rightarrow P * \underset{\sim}{Q} \models \mathfrak{Q})$$

for a cardinal θ.

If κ *is Laver-generically strongly superhuge for* $(\mathfrak{P}, \mathfrak{Q})$, *then, for any cardinal* $\mu > \kappa$ *and* $\mathbb{P} = \mathrm{Col}(\kappa, \mu)$, *we have*

(4.25) $\Vdash_{\mathbb{P}}$ "κ *is strongly generically superhuge for* $\mathfrak{Q}$ ".

Proof. Let $\mathbb{G}$ be a $(\mathsf{V}, \mathbb{P})$-generic filter. For cardinals λ, let $\lambda' = \max\{(\mu^{<\kappa})^+, \lambda, \theta\}$. In $\mathsf{V}[\mathbb{G}]$, let $\mathbb{Q}$ be a poset with $\mathbb{Q} \models \mathfrak{Q}$ and $\mathbb{H}$ a $(\mathsf{V}[\mathbb{G}], \mathbb{Q})$-generic filter such that there are j, $M \subseteq \mathsf{V}[\mathbb{G}][\mathbb{H}]$ satisfying: M is a transitive class in $\mathsf{V}[\mathbb{G}][\mathbb{H}]$; $j : \mathsf{V} \overset{\preccurlyeq}{\to} M$; $crit(j) = \kappa$; $j(\kappa) > \lambda'$; $\mathbb{G}, \mathbb{H} \in M$; and

(4.26) $([M]^{j(\kappa)})^{\mathsf{V}[\mathbb{G}][\mathbb{H}]} \subseteq M$.

Let $\mathbb{P}^* = j(\mathbb{P})$. By elementarity and (4.26), we have $\mathbb{P}^* = \mathrm{Col}(j(\kappa, j(\mu)))$ in $\mathsf{V}[\mathbb{G}][\mathbb{H}]$. Since $\mathbb{P}^*$ is $j(\kappa)$-directed closed (in M or in $\mathsf{V}[\mathbb{G}][\mathbb{H}]$), there is (a master condition) $\mathbb{r} \in \mathbb{P}^*$ with $\mathbb{r} \leq_{\mathbb{P}^*} j(\mathbb{p})$ for all $\mathbb{p} \in \mathbb{G}$. Let $\mathbb{G}^*$ be $(\mathsf{V}[\mathbb{G}][\mathbb{H}], \mathbb{P}^*)$-generic filter with $\mathbb{r} \in \mathbb{G}^*$. Then j is lifted to

(4.27) $\tilde{j} : \mathsf{V}[\mathbb{G}] \overset{\preccurlyeq}{\to} M[\mathbb{G}^*]$; $\underset{\sim}{a}[\mathbb{G}] \mapsto j(\underset{\sim}{a})[\mathbb{G}^*]$

for $\mathbb{P}$-name $\underset{\sim}{a}$. $j \subseteq \tilde{j}$ and hence it is clear that $\tilde{j}$ satisfies: $crit(\tilde{j}) = \kappa$; $\tilde{j}(\kappa) > \lambda$; $\mathbb{G}^* \in M[\mathbb{G}^*]$ since $\underset{\sim}{\mathbb{G}}^* \in M$ by (4.26) and $M \models \mathsf{ZFC}$ where $\underset{\sim}{\mathbb{G}}^*$ is the standard $\mathbb{P}^*$-name of $\mathbb{G}^*$. We also have

(4.28) $([M[\mathbb{G}^*]]^{j(\kappa)})^{\mathsf{V}[\mathbb{G}][\mathbb{H}][\mathbb{G}^*]} \subseteq M[\mathbb{G}^*]$

by Lemma 4.2. This shows that (4.25) holds. □ (Proposition 4.6)

Proposition 4.7. (1) *Let* $\mathfrak{P}$ *be the property "forcing equivalent to a poset of the form* $\mathrm{Col}(\kappa, \mu)$ *for some* μ" *and* $\mathfrak{Q}_\theta$ *is the property "forcing equivalent to a regular sub-poset of the completion of a poset of the form 'generalized Cohen poset* $\times$ $<\theta$*-closed poset' ". If* κ *is strongly Laver-generically supercompact for* $(\mathfrak{P}, \mathfrak{Q}_\theta)$ *for all* $\theta \in \mathsf{Card}$, *then, for any cardinal* μ *and* $\mathbb{P} = \mathrm{Col}(\kappa, \mu)$, *we have*

(4.29) $\Vdash_{\mathbb{P}}$ "$\mathsf{HH}(<\kappa)$ ".

(2) *Let* $\mathfrak{P}$ *be the property "forcing equivalent to a poset of the form* $\mathrm{Col}(\kappa, \mu)$ *for some* μ" *and* $\mathfrak{Q}_\theta$ *is the property "forcing equivalent to a regular sub-poset of the completion of the poset of the form 'ccc poset* $\times$ $<\theta$*-closed poset"'. If* κ *is strongly Laver-generically supercompact for* $(\mathfrak{P}, \mathfrak{Q}_\theta)$ *for all* $\theta \in \mathsf{Card}$, *then, for any cardinal* μ *and* $\mathbb{P} = \mathrm{Col}(\kappa, \mu)$, *we have*

(4.30) $\Vdash_{\mathbb{P}}$ "$\mathsf{SDLS}^{int}_{+}(\mathcal{L}^{\mathsf{PKL}}_{stat}, <\kappa)$ ".

Proof. (1): Suppose that $\mathbb{P} = \mathrm{Col}(\kappa, \mu)$ and $\mathbb{G}$ is a $(\mathsf{V}, \mathbb{P})$-generic filter. In $\mathsf{V}[\mathbb{G}]$, let $X = \langle X, \tau\rangle$ be a non-metrizable topological space such that

(4.31) $\chi(a, X) < \kappa$ for all $a \in X$.

Let $\lambda_0 = |\, X\,|$, $\theta = \lambda = \max\{(2^{\lambda_0})^+, \mu^{<\kappa}\}$ and let $j : V \overset{\preccurlyeq}{\to} M \subseteq \mathsf{V}[\mathbb{G}][\mathbb{H}]$ be such that $crit(j) = \kappa$,

(4.32) $j(\kappa) > \lambda$, $j''\lambda \in M$, and
(4.33) $\left([M]^{\aleph_0}\right)^{\mathsf{V}[\mathbb{G}][\mathbb{H}]} \subseteq M$,

where $\mathbb{H} = \tilde{\mathbb{H}} \cap \mathbb{Q}$ for a $(\mathsf{V}[\mathbb{G}], \tilde{\mathbb{Q}})$-generic filter $\tilde{\mathbb{H}}$ for a poset $\tilde{\mathbb{Q}}$ in $\mathsf{V}[\mathbb{G}]$ of the form

(4.34) $\tilde{\mathbb{Q}}$ $\sim$ generalized Cohen poset $\times$ $<\theta$-closed poset

and $\mathbb{Q} \lessdot \tilde{\mathbb{Q}}$.

Let $\mathbb{P}^* = j(\mathbb{P})$. By elementarity $M \models$ "$\mathbb{P}^*$ is $j(\kappa)$-directed closed". Since $|\,\mathbb{P}\,| \leq \lambda$, $j''\mathbb{G} \in M[\mathbb{G}]$ by (4.32) (see Lemma 2.5 in [Fuchino, Ottenbreit and Sakai[10]]). Since $\lambda < j(\kappa)$ there is $\mathbb{r} \in \mathbb{P}^*$ such that $\mathbb{r} \leq_{\mathbb{P}^*} j(\mathbb{p})$ for all $\mathbb{p} \in \mathbb{G}$. Let $\mathbb{G}^*$ be a $(\mathsf{V}[\mathbb{G}][\tilde{\mathbb{H}}], \mathbb{P}^*)$-generic filter with $\mathbb{r} \in \mathbb{G}^*$. j is then lifted to

(4.35) $$\tilde{j} : \mathsf{V}[\mathbb{G}] \overset{\preccurlyeq}{\to} M[\mathbb{G}^*] \subseteq \mathsf{V}[\mathbb{G}][\mathbb{H}][\mathbb{G}^*] \subseteq \mathsf{V}[\mathbb{G}][\tilde{\mathbb{H}}][\mathbb{G}^*]\,;$$
$$\underset{\sim}{a}[\mathbb{G}] \mapsto j(\underset{\sim}{a})[\mathbb{G}^*].$$

Let $\tau_0 = \{j(O) \cap j''X \,:\, O \in \tau\}$. By (4.31) and since κ is the critical point of j, we have

(4.36) $M \models$ "$\langle j''X, \tau_0\rangle$ is a subspace of $\langle j(X), j(\tau)\rangle$".

We also have

(4.37) $M[\mathbb{G}^*] \models$ "$\langle j''X, \tau_0\rangle$ is homeomorphic to $\langle X, \tau\rangle$".

Hence the same property holds in $\mathsf{V}[\mathbb{G}][\tilde{\mathbb{H}}][\mathbb{G}^*]$.

Now generalized Cohen poset part of $\tilde{\mathbb{H}}$ preserve the non-metrizability of $\langle X, \tau\rangle$ by Theorem 1.1. By the $<\theta$-closed part of $\tilde{\mathbb{H}}$ no new metric on X is added. Hence $\mathsf{V}[\mathbb{G}][\tilde{\mathbb{H}}] \models$ " $\langle X, \tau\rangle$ is non-metrizable". It follows that $M \models$ " $\langle X, \tau\rangle$ is non-metrizable" and hence by $<\lambda$-closedness of $\mathbb{P}^*$, it follows that $M[\mathbb{G}^*] \models$ "$\langle X, \tau\rangle$ is non-metrizable". Thus by (4.37), $M[\mathbb{G}^*] \models$ " $\langle \tilde{j}''X, \tau_0\rangle$ is non-metrizable". Thus

(4.38) $M[\mathbb{G}^*] \models$ "$\tilde{j}(X)$ has a non-metrizable subspace Y of cardinality $< \tilde{j}(\kappa)$".

By elementarity of $\tilde{j}$ it follows that

(4.39) $\mathsf{V}[\mathbb{G}] \models$ "X has a non-metrizable subspace Y of cardinality $< \kappa$".

(2): The proof is done similarly to (1), by using Lemma 3.1 in place of Theorem 1.1. ❑ (Proposition 4.7)

Lemma 4.8. *Let $\mathfrak{Q}_\theta$ for a cardinal θ be as in* Proposition 4.7, (2) *and assume that κ is strongly generically superhuge for $\mathfrak{Q}_\theta$ for all $\theta \in$ Card. Then, for any $\lambda \geq \kappa$, $\mathcal{P}_\kappa(\lambda)$ carries a σ-saturated normal ideal.*

Proof. Let $\lambda \geq \kappa$ and let $\mathbb{Q} \lessdot RO(\mathbb{S} \times \mathbb{T})$ be such that $\mathbb{S}$ is $<\left(2^{2^{(\lambda^{<\kappa})}}\right)^+$-closed poset, $\mathbb{T}$ is ccc, and there is a $(\mathsf{V}, \mathbb{Q})$-generic filter $\mathbb{H}$ and j, $M \subseteq \mathsf{V}[\mathbb{H}]$ such that $j : \mathsf{V} \overset{\preccurlyeq}{\to} M$, $crit(j) = \kappa$, $j(\kappa) > \lambda$, $j''\lambda \in M$ and $([M]^{j(\kappa)})^{\mathsf{V}[\mathbb{H}]} \subseteq M$. Let $\mathbb{K}$ be a $(\mathsf{V}, \mathbb{S})$-generic filter and $\mathbb{L}$ a $(\mathsf{V}[\mathbb{K}], \mathbb{T})$-generic filter such that $\mathsf{V}[\mathbb{H}] \subseteq \mathsf{V}[\mathbb{K}][\mathbb{L}]$. In $\mathsf{V}[\mathbb{K}][\mathbb{L}]$,

(4.40) $\mathcal{I} = \{X \in \left(\mathcal{P}(\mathcal{P}_\kappa(\lambda)^{\mathsf{V}})\right)^{\mathsf{V}} : j''\lambda \notin j(X)\}$

is a V-normal ideal. Since $\mathsf{V}[\mathbb{K}] \models$ "$\mathbb{T}$ is ccc", it follows that, in $\mathsf{V}[\mathbb{K}]$, $\mathcal{I}' = \{X \in \left(\mathcal{P}(\mathcal{P}_\kappa(\lambda)^{\mathsf{V}})\right)^{\mathsf{V}} : \Vdash_{\mathbb{T}}$ "$\check{X} \in \underset{\sim}{\mathcal{I}}$"$\}$ is a σ-saturated V-normal filter for $\mathbb{T}$-name $\underset{\sim}{\mathcal{I}}$ of $\mathcal{I}$. Now, by the closedness of $\mathbb{S}$, $\mathcal{I}' \in \mathsf{V}$ and $\mathcal{I}'$ is a σ-saturated normal filter in V. ❑ (Lemma 4.8)

Proposition 4.9. *Let $\mathfrak{P}$ and $\mathfrak{Q}_\theta$ for a cardinal θ be as in* Proposition 4.7, (2). *If κ is strongly Laver-generically superhuge for $(\mathfrak{P}, \mathfrak{Q}_\theta)$ for all cardinal θ, then, or any $\lambda \geq \kappa$, and $\mathbb{P} = \mathrm{Col}(\kappa, \lambda)$,*

(4.41) $\Vdash_{\mathbb{P}}$ "*$\mathcal{P}_\kappa(\lambda)$ carries a σ-saturated normal ideal*".

Proof. By Proposition 4.6 and Lemma 4.8. ❑ (Corollary 4.9)

5. Mixed support iteration

The construction of the mixed support iteration we give here is similar to the one given in [Krueger [15,16]]. Nevertheless, we will examine the details of our construction in the following, since there are a couple of points organized differently from [Krueger [15,16]].

In this section, κ is always a fixed supercompact cardinal and $f : \kappa \to \mathsf{V}_\kappa$ is a Laver function, i.e. a function satisfying:

(5.1) for any set a and any $\lambda \geq \kappa$, there is $j : V \overset{\preccurlyeq}{\rightarrow} M$ such that $crit(j) = \kappa$, $j(\kappa) > \lambda$, $[M]^\lambda \subseteq M$ and $j(f)(\kappa) = a$

(see e.g. Theorem 20.21 in [Jech[13]]).

Let $\overline{f} : \kappa \to \kappa$ be defined by

$$\overline{f}(\alpha) = |\, trcl(f(\alpha))\,| \text{ for } \alpha < \kappa. \tag{5.2}$$

Let

$$S = \{\alpha < \kappa : \alpha \text{ is a strongly Mahlo cardinal closed with respect to } \overline{f}\ \}, \text{ and let} \tag{5.3}$$

$$T = \kappa \setminus S. \tag{5.4}$$

Let $\nu : \kappa \to \kappa$ be the mapping defined by

$$\nu(\alpha) = \min(S \setminus (\alpha + 1)) \text{ for } \alpha \in \kappa. \tag{5.5}$$

We treat iterations here as in [Jech[13]] such that elements of αth step $\mathbb{P}_\alpha$ of an iteration $\langle \mathbb{P}_\alpha, \underset{\sim}{\mathbb{Q}}_\beta : \alpha \leq \kappa, \beta < \kappa \rangle$ are sequences of length α.

Let $\langle \mathbb{O}_\alpha, \underset{\sim}{\mathbb{R}}_\beta : \alpha \leq \kappa, \beta < \kappa \rangle$ be a finite support iteration of ccc posets which will be further specified later. This preparatory iteration should satisfy the following conditions:

$$\underset{\sim}{\mathbb{R}}_\alpha \in V_{\nu(\alpha)}, \text{ and} \tag{5.6}$$

$$\Vdash_{\mathbb{O}_\alpha} \text{“}\underset{\sim}{\mathbb{R}}_\alpha \equiv \{\mathbb{1}_{\underset{\sim}{\mathbb{R}}_\alpha}\}\text{”, for all } \alpha \in S. \tag{5.7}$$

We denote the canonical embeddings of $\mathbb{O}_\alpha$ into $\mathbb{O}_\beta$ for $\alpha \leq \beta \leq \kappa$ by $i^*_{\alpha,\beta}$. Thus, $i^*_{\alpha,\beta}$ is the mapping defined by $i^*_{\alpha,\beta}(\mathbb{p}) = \mathbb{p} \cup \vec{\mathbb{1}}_{\alpha,\beta}$, where $\vec{\mathbb{1}}_{\alpha,\beta}$ is the function g on $\beta \setminus \alpha$ with $g(\xi) = \mathbb{1}_{\underset{\sim}{\mathbb{R}}_\xi}$ for all $\xi \in \beta \setminus \alpha$.

$\langle \mathbb{P}_\alpha, \underset{\sim}{\mathbb{Q}}_\beta : \alpha \leq \kappa, \beta < \kappa \rangle$ is our final iteration which is specified once the preparatory iteration $\langle \mathbb{O}_\alpha, \underset{\sim}{\mathbb{R}}_\beta : \alpha \leq \kappa, \beta < \kappa \rangle$ is fixed. The iteration $\langle \mathbb{P}_\alpha, \underset{\sim}{\mathbb{Q}}_\beta : \alpha \leq \kappa, \beta < \kappa \rangle$ is defined recursively in (A) and (B) below, together with the commutative systems of complete embeddings $\iota_\alpha : \mathbb{O}_\alpha \to \mathbb{P}_\alpha$, and $i_{\alpha,\beta} : \mathbb{P}_\alpha \to \mathbb{P}_\beta$ for $\alpha \leq \beta \leq \kappa$ which should satisfy

$$i_{\alpha,\alpha} = id_{\mathbb{P}_\alpha} \text{ for } \alpha \leq \kappa; \tag{5.8}$$

$$\iota_\beta \circ i^*_{\alpha,\beta} = i_{\alpha,\beta} \circ \iota_\alpha, \text{ and;} \tag{5.9}$$

$$i_{\beta,\gamma} \circ i_{\alpha,\beta} = i_{\alpha,\gamma} \text{ for } \alpha \leq \beta \leq \gamma \leq \kappa; \tag{5.10}$$

(5.11) $\text{supp}(\iota_\beta(\mathbb{o})) = \text{supp}(\mathbb{o})$, where $\text{supp}(\cdot)$ is defined as in (5.17) below, and

(5.12) $\iota_\alpha(\mathbb{o} \restriction \alpha) = \iota_\beta(\mathbb{o}) \restriction \alpha$ for $\alpha < \beta \leq \kappa$ and $\mathbb{o} \in \mathbb{O}_\beta$.

We define now the *Easton-type mixed support* of the iteration as a sequence $\langle \mathcal{I}_\alpha : \alpha < \kappa \rangle$ of ideals where each $\mathcal{I}_\alpha$ for $\alpha \leq \kappa$ is an ideal over α.

(5.13) $\mathcal{I}_{\alpha+1} = \mathcal{I}_\alpha \cup \{s \cup \{\alpha\} : s \in \mathcal{I}_\alpha\}$ for all $\alpha < \kappa$;

(5.14) If $\gamma < \kappa$ is a limit ordinal but not a regular cardinal, then $\mathcal{I}_\gamma = \{s \subseteq \gamma : s \cap \alpha \in \mathcal{I}_\alpha$ for all $\alpha < \gamma$ and $s \cap T$ is bounded in $\gamma\}$;

(5.15) If $\gamma \leq \kappa$ is a regular cardinal, then $\mathcal{I}_\gamma = \{s \subseteq \gamma : s \cap \alpha \in \mathcal{I}_\alpha$ for all $\alpha < \gamma$ and s is bounded in $\gamma\}$.

The following is easy to prove by induction on $\alpha \leq \kappa$:

Lemma 5.1. (1) *$\mathcal{I}_\alpha$ is an ideal over α with $\{\{\beta\} : \beta < \alpha\} \subseteq \mathcal{I}_\alpha$ for all $\alpha \leq \kappa$.*

(2) *For all $\alpha \leq \kappa$, $s \in \mathcal{I}_\alpha \Leftrightarrow s \cap T$ is finite and $|s \cap \mu| < \mu$ for all regular infinite cardinal $\mu \leq \alpha$.* □

Now we are ready to define the iteration $\langle \mathbb{P}_\alpha, \underset{\sim}{\mathbb{Q}}_\beta : \alpha \leq \kappa, \beta < \kappa \rangle$ in the following (A) and (B):

(A) If $\langle \mathbb{P}_\alpha, \underset{\sim}{\mathbb{Q}}_\beta : \alpha < \gamma, \beta < \gamma \rangle$ has been defined for a limit $\gamma \leq \kappa$, let

(5.16) $\mathbb{P}_\gamma = \{\mathbb{p} : \mathbb{p}$ is a sequence of length γ, $\mathbb{p} \restriction \alpha \in \mathbb{P}_\alpha$ for all $\alpha < \gamma$, and $\operatorname{supp}(\mathbb{p}) \in \mathcal{I}_\gamma \}$

where

(5.17) $\operatorname{supp}(\mathbb{p}) = \{\alpha < \gamma : \mathbb{p}(\alpha) \neq \mathbb{1}_{\underset{\sim}{\mathbb{Q}}_\alpha}\}$.

For $\mathbb{p}, \mathbb{q} \in \mathbb{P}_\gamma$,

(5.18) $\mathbb{p} \leq_{\mathbb{P}_\gamma} \mathbb{q} \Leftrightarrow \mathbb{p} \restriction \delta \leq_{\mathbb{P}_\delta} \mathbb{q} \restriction \delta$ for all $\delta < \gamma$.

We assume that the complete embeddings $\iota_\alpha : \mathbb{O}_\alpha \to \mathbb{P}_\alpha$, $\alpha < \gamma$ have been defined such that (5.11) and (5.12) hold for all $\alpha \leq \beta < \gamma$. The mapping ι_γ is then defined by:

(5.19) $\iota_\gamma : \mathbb{O}_\gamma \to \mathbb{P}_\gamma;\ \mathbb{o} \mapsto \bigcup_{\delta<\gamma} \iota_\delta(\mathbb{o} \restriction \delta)$.

For $\delta \leq \gamma$, let $\vec{\mathbb{1}}_{\delta,\gamma} = \{\langle \alpha, \mathbb{1}_{\underset{\sim}{\mathbb{Q}}_\alpha} \rangle : \delta \leq \alpha < \gamma\}$ as before and let $i_{\delta,\gamma}$ be defined by

(5.20) $i_{\delta,\gamma} : \mathbb{P}_\delta \to \mathbb{P}_\gamma;\ \mathbb{p} \mapsto \mathbb{p} \cup \vec{\mathbb{1}}_{\delta,\gamma}$.

It is easy to check that ι_γ and $i_{\delta,\gamma}$, $\gamma \leq \delta$ are complete embeddings and (5.8) $\sim$ (5.11) hold for all indices $\leq \gamma$.

(B) Now suppose that $\langle \mathbb{P}_\alpha, \underset{\sim}{\mathbb{Q}}_\beta : \alpha \leq \gamma, \beta < \gamma \rangle$, $\langle \iota_\beta : \beta < \gamma \rangle$ and $\langle i_{\alpha,\beta} : \alpha \leq \beta \leq \gamma \rangle$ have been defined for some $\gamma < \kappa$.

(a) If $\gamma \in S$ and

(5.21) $\quad f(\gamma) = \langle \mu, \theta, R \rangle$ for some cardinals μ, $\theta > \gamma$ and a set R,[c]

then let

(5.22) $\quad \underset{\sim}{\mathbb{Q}}_\gamma = (\mathrm{Col}(\gamma, \mu))^{\bullet}_{\mathbb{P}_\gamma}$, and

(5.23) $\quad \mathbb{P}_{\gamma+1} = \{ \mathbb{p} \cup \{ \langle \gamma, \underset{\sim}{\mathbb{q}} \rangle \} : \mathbb{p} \in \mathbb{P}_\gamma, \underset{\sim}{\mathbb{q}}$ is a canonical $\mathbb{P}_\gamma$-name[d] such that $\Vdash_{\mathbb{P}_\gamma}$ "$\underset{\sim}{\mathbb{q}} \,\varepsilon\, \underset{\sim}{\mathbb{Q}}_\gamma$"$\}$.

For $\mathbb{p}_0 \cup \{ \langle \gamma, \underset{\sim}{\mathbb{q}}_0 \rangle \}$, $\mathbb{p}_1 \cup \{ \langle \gamma, \underset{\sim}{\mathbb{q}}_1 \rangle \} \in \mathbb{P}_{\gamma+1}$,

(5.24) $\quad \mathbb{p}_0 \cup \{ \langle \gamma, \underset{\sim}{\mathbb{q}}_0 \rangle \} \leq_{\mathbb{P}_{\gamma+1}} \mathbb{p}_1 \cup \{ \langle \gamma, \underset{\sim}{\mathbb{q}}_1 \rangle \} \Leftrightarrow \mathbb{p}_0 \leq_{\mathbb{P}_\gamma} \mathbb{p}_1$ and $\mathbb{p}_0 \Vdash_{\mathbb{P}_\gamma}$ "$\underset{\sim}{\mathbb{q}}_0 \leq_{\underset{\sim}{\mathbb{Q}}_\gamma} \underset{\sim}{\mathbb{q}}_1$".

For $\mathbb{o} \in \mathbb{O}_{\gamma+1}$, let

(5.25) $\quad \iota_{\gamma+1}(\mathbb{o}) = \iota_\gamma(\mathbb{o} \restriction \gamma) \cup \{ \langle \gamma, \mathbb{1}_{\underset{\sim}{\mathbb{Q}}_\gamma} \rangle \}$,

and, for $\alpha \leq \gamma$ and $\mathbb{p} \in \mathbb{P}_\alpha$, let

(5.26) $\quad i_{\alpha,\gamma+1}(\mathbb{p}) = i_{\alpha,\gamma}(\mathbb{p}) \cup \{ \langle \gamma, \mathbb{1}_{\underset{\sim}{\mathbb{Q}}_\gamma} \rangle \}$.

(b) If $\gamma \in S$ but (a) does not hold, then let $\underset{\sim}{\mathbb{Q}}_\gamma$ be a $\mathbb{P}_\gamma$-name of trivial forcing and the rest is treated just as in the case (a).

In both of the cases (a) and (b), it is clear that the defined mappings are complete embeddings and satisfy (5.8) $\sim$ (5.11).

(c) If $\gamma \notin S$, then let $\underset{\sim}{\mathbb{Q}}_\gamma$ be the $\mathbb{P}_\gamma$-name $\iota_\gamma(\underset{\sim}{\mathbb{R}}_\gamma)$ and[e]

(5.27) $\quad \mathbb{P}_{\gamma+1} = \{ \mathbb{p} \cup \{ \langle \gamma, \iota_\gamma(\underset{\sim}{\mathbb{r}}) \rangle \} : \mathbb{p} \in \mathbb{P}_\gamma$, $\underset{\sim}{\mathbb{r}}$ is a canonical $\mathbb{O}_\gamma$-name such that $\Vdash_{\mathbb{O}_\gamma}$ "$\underset{\sim}{\mathbb{r}} \,\varepsilon\, \underset{\sim}{\mathbb{R}}_\gamma$"$\}$.

---

[c] At the moment, R does not play any role. This component is added here so that we can later modify the construction.

[d] Adopting the terminology of [Cummings[4]], we call a $\mathbb{P}$-name $\underset{\sim}{a}$ is a *canonical* $\mathbb{P}$*-name* if, for any $\mathbb{P}$-name $\underset{\sim}{b}$ with $\Vdash_{\mathbb{P}}$ "$\underset{\sim}{a} \equiv \underset{\sim}{b}$", we have $|\,\mathit{trcl}(\underset{\sim}{a})\,| \leq |\,\mathit{trcl}(\underset{\sim}{b})\,|$.

[e] With ι_γ, we also denote the embedding of $V^{\mathbb{O}_\gamma}$ into $V^{\mathbb{P}_\gamma}$ canonically induced by ι_γ.

For $\mathbb{p}_0 \cup \{\langle \gamma, \iota_\gamma(\underset{\sim}{\mathbb{r}}_0)\rangle\}$, $\mathbb{p}_1 \cup \{\langle \gamma, \iota_\gamma(\underset{\sim}{\mathbb{r}}_1)\rangle\} \in \mathbb{P}_{\gamma+1}$,

$$(5.28)\quad \mathbb{p}_0 \cup \{\langle \gamma, \iota_\gamma(\underset{\sim}{\mathbb{r}}_0)\rangle\} \ \leq_{\mathbb{P}_{\gamma+1}} \ \mathbb{p}_1 \cup \{\langle \gamma, \iota_\gamma(\underset{\sim}{\mathbb{r}}_1)\rangle\}$$
$$\Leftrightarrow \ \mathbb{p}_0 \leq_{\mathbb{P}_\gamma} \mathbb{p}_1 \text{ and there is } \mathbb{o} \in \mathbb{O}_\gamma \text{ such that } \mathbb{p}_0 \leq_{\mathbb{P}_\gamma} \iota_\gamma(\mathbb{o}) \text{ and } \mathbb{o} \Vdash_{\mathbb{O}_\gamma} \text{“} \underset{\sim}{\mathbb{r}}_0 \leq_{\underset{\sim}{\mathbb{R}}_\gamma} \underset{\sim}{\mathbb{r}}_1 \text{”}.$$

For $\mathbb{o} \in \mathbb{O}_{\gamma+1}$, let

$$(5.29)\quad \iota_{\gamma+1}(\mathbb{o}) = \iota_\gamma(\mathbb{o} \restriction \gamma) \cup \{\langle \gamma, \iota_\gamma(\mathbb{o}(\gamma))\rangle\},$$

and, for $\alpha \leq \gamma$ and $\mathbb{p} \in \mathbb{P}_\alpha$, let

$$(5.30)\quad i_{\alpha,\gamma+1}(\mathbb{p}) = i_{\alpha,\gamma}(\mathbb{p}) \cup \{\langle \gamma, \mathbb{1}_{\underset{\sim}{\mathbb{Q}}_\gamma}\rangle\}.$$

Also in this case, the mapping introduced are complete embeddings and (5.8) ∼ (5.11) are satisfied.

This finishes the construction of our Easton-type mixed support iteration.

The following three lemmas can be proved easily with the standard argument in the order as we present them here.

Lemma 5.2. *For an ordinal $\gamma \leq \kappa$ and a γ-sequence $\mathbb{p}$,*

$$\begin{aligned} \mathbb{p} \in \mathbb{P}_\gamma \ \Leftrightarrow \ & \Vdash_{\mathbb{P}_\xi} \text{“}\mathbb{p}(\xi) \,\varepsilon\, \underset{\sim}{\mathbb{Q}}_\xi\text{” for all } \xi \in S \cap \gamma, \\ & \mathbb{p}(\xi) = \iota_\xi(\underset{\sim}{\mathbb{r}}) \text{ for a canonical } \mathbb{O}_\gamma\text{-name } \underset{\sim}{\mathbb{r}} \text{ with} \\ & \Vdash_{\mathbb{O}_\xi} \text{“}\underset{\sim}{\mathbb{r}} \,\varepsilon\, \underset{\sim}{\mathbb{R}}_\xi\text{” for all } \xi \in T \cap \gamma, \text{ and} \\ & \mathrm{supp}(\mathbb{p}) = \{\xi < \gamma : \mathbb{p}(\xi) \not\equiv \mathbb{1}_{\underset{\sim}{\mathbb{Q}}_\xi}\} \in \mathcal{I}_\gamma. \end{aligned}$$ □

Lemma 5.3. *For $\delta \leq \gamma \leq \kappa$, $\mathbb{p}_0$, $\mathbb{p}_1 \in \mathbb{P}_\gamma$ and $s \subseteq \gamma$, if $\mathrm{supp}(\mathbb{p}_0) \subseteq \delta \cup s$, $\mathrm{supp}(\mathbb{p}_1) \subseteq \delta \cup (\gamma \setminus s)$ and $\mathbb{p}_0 \restriction \delta \leq_{\mathbb{P}_\delta} \mathbb{p}_1 \restriction \delta$, then*

$$\begin{aligned} (5.31)\quad \mathbb{p}_2 &= (\mathbb{p}_0 \restriction (\delta \cup s)) \cup (\mathbb{p}_1 \restriction (\gamma \setminus (\delta \cup s))) \\ &\big(= (\mathbb{p}_0 \restriction \mathrm{supp}(\mathbb{p}_0)) \cup (\mathbb{p}_1 \restriction (\gamma \setminus \mathrm{supp}(\mathbb{p}_0))) \big) \in \mathbb{P}_\gamma \end{aligned}$$

and $\mathbb{p}_2$ is a maximal element of $\mathbb{P}_\gamma$ below $\mathbb{p}_0$ and $\mathbb{p}_1$ with respect to $\leq_{\mathbb{P}_\gamma}$.

Proof. By induction on γ with $\delta \leq \gamma \leq \kappa$. □ (Lemma 5.3)

Note that, in the Lemma above, we are talking about “a” maximal element since $\leq_{\mathbb{P}_\gamma}$ is merely a pre-ordering in general.

The following can be proved applying the Pressing-down Lemma and Lemma 5.3 above. Note that, for $\alpha \in S \cup \{\kappa\}$, $R = \{\beta < \alpha : \mathbb{P}_\beta$ is a direct limit of $\langle \mathbb{P}_\xi : \xi < \alpha \rangle\}$ is a stationary subset of α by (5.3) and (5.15).

Lemma 5.4. *For $\nu \in S \cup \{\kappa\}$, we have $|\mathbb{P}_\mu| < \nu$ for all $\mu < \nu$, $\mathbb{P}_\nu \subseteq V_\nu$ and $\mathbb{P}_\nu$ has the ν-cc.* □

For posets $\mathbb{P}$, $\mathbb{Q}$, a mapping $p : \mathbb{Q} \to \mathbb{P}$ is said to be a *projection* if

(5.32) $p(\mathbb{1}_\mathbb{Q}) = \mathbb{1}_\mathbb{P}$;

(5.33) p is order-preserving; and

(5.34) for any $\mathbb{p} \in \mathbb{P}$ and $\mathbb{q} \in \mathbb{Q}$, if $\mathbb{p} \leq_\mathbb{P} p(\mathbb{q})$, then there is $\mathbb{q}' \in \mathbb{Q}$ such that $\mathbb{q}' \leq_\mathbb{Q} \mathbb{q}$ and $p(\mathbb{q}') \leq_\mathbb{P} \mathbb{p}$.

The following is standard and also easy to check:

Lemma 5.5. *Suppose that $\mathbb{P}$, $\mathbb{Q}$ are posets and $p : \mathbb{Q} \to \mathbb{P}$ is a projection.*

(1) *If $\mathbb{H}$ is a $(\mathsf{V}, \mathbb{Q})$-generic filter, then $p''\mathbb{H}$ generates a $(\mathsf{V}, \mathbb{P})$-generic filter.*

(2) *If $\mathbb{G}$ is a $(V, \mathbb{P})$-generic filter, then letting*

(5.35) $\mathbb{Q}/\mathbb{G} = \{\mathbb{q} \in \mathbb{Q} : p(\mathbb{q}) \in \mathbb{G}\}$

be poset with the pre-ordering $\leq_\mathbb{Q}$ restricted to it, any $(\mathsf{V}[\mathbb{G}], \mathbb{Q}/\mathbb{G})$-generic filter $\mathbb{H}$ is a $(\mathsf{V}, \mathbb{Q})$-generic filer with $p''\mathbb{H} \subseteq \mathbb{G}$. □

Suppose that $\langle \mathbb{P}_\alpha, \underset{\sim}{\mathbb{Q}}_\beta : \alpha \leq \kappa, \beta < \kappa \rangle$ is an Easton-type mixed support iteration with the Laver-function $f : \kappa \to V_\kappa$ and S as above over a finite support iteration $\langle \mathbb{O}_\alpha, \underset{\sim}{\mathbb{R}}_\beta : \alpha \leq \kappa, \beta < \kappa \rangle$.

Note that, for $\alpha \leq \beta \leq \kappa$,

(5.36) $p_{\beta,\alpha} : \mathbb{P}_\beta \to \mathbb{P}_\alpha$; $\mathbb{q} \mapsto \mathbb{q} \restriction \alpha$ is a projection, and

(5.37) $p_{\beta,\alpha} \circ i_{\alpha,\beta} = id_{\mathbb{P}_\alpha}$.

For $\delta_0 < \kappa$, let $\mathbb{G}_{\delta_0}$ be a $(\mathsf{V}, \mathbb{P}_{\delta_0})$-generic filter. Working in $\mathsf{V}[\mathbb{G}_{\delta_0}]$, let $\delta_0 \leq \gamma \leq \kappa$, and let

(5.38) $\mathbb{P}_\gamma/\mathbb{G}_{\delta_0} = \{\mathbb{p} \in \mathbb{P}_\gamma : \mathbb{p} \restriction \delta_0 \in \mathbb{G}_{\delta_0}\}$

be the poset with the pre-ordering $\leq_{\mathbb{P}_\gamma}$ restricted to $\mathbb{P}_\gamma/\mathbb{G}_{\delta_0}$ and with the designated maximal element $\mathbb{1}_{\mathbb{P}_\gamma/\mathbb{G}_{\delta_0}} = \mathbb{1}_{\mathbb{P}_\gamma}$.

Lemma 5.6. (1) *A $(\mathsf{V}[\mathbb{G}_{\delta_0}], \mathbb{P}_\gamma/\mathbb{G}_{\delta_0})$-generic filter $\mathbb{H}$ is also a $(\mathsf{V}, \mathbb{P}_\gamma)$-generic filter with $i_{\delta_0,\gamma}{}''\mathbb{G}_{\delta_0} \subseteq \mathbb{H}$.*

(2) *If $\mathbb{H}$ is a $(\mathsf{V}, \mathbb{P}_\gamma)$-generic filter with $i_{\delta_0,\gamma}{}''\mathbb{G} \subseteq \mathbb{H}$, then $\mathbb{H}$ is a $(\mathsf{V}[\mathbb{G}_{\delta_0}], \mathbb{P}_\gamma/\mathbb{G}_{\delta_0})$-generic filter.*

Proof. By (5.36), (5.37) and Lemma 5.5. □ (Lemma 5.6)

It is well-known that projections and complete embeddings are two interchangeable notions for cBa posets[f]:

Lemma 5.7. *For cBa posets $\mathbb{P}$ and $\mathbb{Q}$, there is a complete embedding $i : \mathbb{P} \to \mathbb{Q}$ if and only if there is a projection $p : \mathbb{Q} \to \mathbb{P}$.*

For cBa posets complete embeddings are injections and projections are surjections. □

For the analysis of the structure of the iteration $\langle \mathbb{P}_\alpha, \underset{\sim}{\mathbb{Q}}_\beta : \alpha \leq \kappa, \beta < \kappa \rangle$, the following alternative treatment of the quotient $\mathbb{P}_\gamma / \mathbb{G}_{\delta_0}$ proves often to be more appropriate.

For $\mathbb{p}_0, \mathbb{p}_1 \in \mathbb{P}_\gamma$, with $\mathrm{supp}(\mathbb{p}_0) \cap \mathrm{supp}(\mathbb{p}_1) = \emptyset$, we denote with $\mathbb{p}_0 \wedge_{\mathbb{P}_\gamma} \mathbb{p}_1$, the element $\mathbb{P}_\gamma$ defined by:

$$(5.39) \quad \mathbb{p}_0 \wedge_{\mathbb{P}_\gamma} \mathbb{p}_1 = \mathbb{p}_0 \restriction \mathrm{supp}(\mathbb{p}_0) \cup \mathbb{p}_1 \restriction (\gamma \setminus \mathrm{supp}(\mathbb{p}_0)).$$

Since it causes no confusion, we drop the subscript $\mathbb{P}_\gamma$ in this notation and simply write $\mathbb{p}_0 \wedge \mathbb{p}_1$ in place of $\mathbb{p}_0 \wedge_{\mathbb{P}_\gamma} \mathbb{p}_1$.

Suppose $\delta_0 < \gamma \leq \kappa$ and $\mathbb{G}_{\delta_0}$ is a $(\mathsf{V}, \mathbb{P}_{\delta_0})$-generic filter. In $\mathsf{V}[\mathbb{G}_{\delta_0}]$, let $\mathbb{P}_\gamma \mid \mathbb{G}_{\delta_0} = \{\mathbb{p} \in \mathbb{P}_\gamma : \mathrm{supp}(\mathbb{p}) \subseteq \gamma \setminus \delta_0\}$ be the poset with the pre-ordering $\leq_{\mathbb{P}_\gamma \mid \mathbb{G}_{\delta_0}}$ defined by

$$(5.40) \quad \mathbb{q}_0 \leq_{\mathbb{P}_\gamma \mid \mathbb{G}_{\delta_0}} \mathbb{q}_1 \;\Leftrightarrow\; i_{\delta_0,\gamma}(\mathbb{p}) \wedge_{\mathbb{P}_\gamma} \mathbb{q}_0 \leq_{\mathbb{P}_\gamma} i_{\delta_0,\gamma}(\mathbb{p}) \wedge_{\mathbb{P}_\gamma} \mathbb{q}_1 \text{ for some } \mathbb{p} \in \mathbb{G}_{\delta_0},$$

for $\mathbb{q}_0, \mathbb{q}_1 \in \mathbb{P}_\gamma \mid \mathbb{G}_{\delta_0}$, and with the designated maximal element $\mathbb{1}_{\mathbb{P}_\gamma \mid \mathbb{G}_{\delta_0}} = \mathbb{1}_{\mathbb{P}_\gamma}$.

Note that, for $\mathbb{q}_0, \mathbb{q}_1 \in \mathbb{P}_\gamma \mid \mathbb{G}_{\delta_0}$,

$$(5.41) \quad \mathbb{q}_0 \leq_{\mathbb{P}_\gamma} \mathbb{q}_1 \text{ implies } \mathbb{q}_0 \leq_{\mathbb{P}_\gamma \mid \mathbb{G}_{\delta_0}} \mathbb{q}_1,$$

since $\mathbb{1}_{\mathbb{P}_{\delta_0}} \in \mathbb{G}_{\delta_0}$.

In the following, just for convenience, we shall often misuse the notation and write instead of $i_{\delta_0,\gamma}(\mathbb{p}) \wedge_{\mathbb{P}_\gamma} \mathbb{q}_0$ etc. simply $\mathbb{p} \wedge \mathbb{q}_0$ etc. The following Lemma is also formulated in this sloppy handling of the notation.

Lemma 5.8. (1) *For $\mathbb{p} \in \mathbb{P}_{\delta_0}$ and $\mathbb{q}_0, \mathbb{q}_1 \in \mathbb{P}_\gamma \mid \mathbb{G}_{\delta_0}$ If $\mathbb{p} \wedge \mathbb{q}_0 \leq_{\mathbb{P}_\gamma} \mathbb{p} \wedge \mathbb{q}_1$, then for $\mathbb{p}' \leq_{\mathbb{P}_{\delta_0}} \mathbb{p}$, we have $\mathbb{p}' \wedge \mathbb{q}_0 \leq_{\mathbb{P}_\gamma} \mathbb{p}' \wedge \mathbb{q}_1 \leq_{\mathbb{P}_\gamma} \mathbb{p} \wedge \mathbb{q}_1$.*

[f]We call a poset $\mathbb{P} = \langle \mathbb{P}, \leq_{\mathbb{P}} \rangle$ a *cBa poset* if (the underlying set) $\mathbb{P}$ of the poset coincides with the positive elements of a complete Boolean algebra and $\leq_{\mathbb{P}}$ coincides with the ordering of the complete Boolean algebra.

(2) *For* $q_0, q_1 \in \mathbb{P}_\gamma \,|\, \mathbb{G}_{\delta_0}$ *with* $\operatorname{supp}(q_0) \cap \operatorname{supp}(q_1) = \emptyset$, $q_0 \wedge q_1$ *is a join of* q_0 *and* q_1 *both with respect to* $\leq_{\mathbb{P}_\gamma}$ *and with respect to* $\leq_{\mathbb{P}\,|\,\mathbb{G}_{\delta_0}}$.

Proof. (1): By Lemma 5.3.

(2): $q_0 \wedge q_1$ is a join of q_0 and q_1 with respect to $\leq_{\mathbb{P}_\gamma}$ by Lemma 5.3. By (5.41), it follows that $q_0 \wedge q_1 \leq_{\mathbb{P}_\gamma \,|\, \mathbb{G}_{\delta_0}} q_0, q_1$. Suppose now that $r \leq_{\mathbb{P}_\gamma \,|\, \mathbb{G}_{\delta_0}} q_0, q_1$. Then there are $s_0, s_1 \in \mathbb{G}_{\delta_0}$ such that $s_0 \wedge r \leq_{\mathbb{P}_\gamma} s_0 \wedge q_0$ and $s_1 \wedge r \leq_{\mathbb{P}_\gamma} s_1 \wedge q_1$. Let $s_2 \in \mathbb{G}_{\delta_0}$ be such that $s_2 \leq_{\mathbb{P}_{\delta_0}} s_0, s_1$. Then, by (1), we have $s_2 \wedge r \leq_{\mathbb{P}_\gamma} s_2 \wedge q_0, s_2 \wedge q_1$. By Lemma 5.3, it follows that $s_2 \wedge r \leq_{\mathbb{P}_\gamma} s_2 \wedge (q_0 \wedge q_1)$. Thus $r \leq_{\mathbb{P}_\gamma \,|\, \mathbb{G}_{\delta_0}} q_0 \wedge q_1$. □ (Lemma 5.8)

For $\gamma \leq \kappa$, $p \in \mathbb{P}_\gamma$ and $X \subseteq \kappa$, let $p \downharpoonright X$ be the condition $r \in \mathbb{P}_\gamma$ defined by

$$(5.42) \quad r(\alpha) = \begin{cases} p(\alpha), \text{ if } \alpha \in X; \\ \mathbb{1}_{\underset{\sim}{\mathbb{Q}}_\alpha}, \text{ otherwise} \end{cases}$$

for all $\alpha < \gamma$.

Since $\operatorname{supp}(p \downharpoonright X) \subseteq \operatorname{supp}(p)$, we have $p \downharpoonright X \in \mathbb{P}_\gamma$ by Lemma 5.2. By definition, it is also clear that $p \leq_{\mathbb{P}_\gamma} p \downharpoonright X$.

For $X \subseteq \gamma$ and $P \subseteq \mathbb{P}_\gamma$, let us write

$$(5.43) \quad P \downharpoonright X = \{p \downharpoonright X \,:\, p \in P\}.$$

Note that the underlying set of $\mathbb{P}_\gamma \,|\, \mathbb{G}_{\delta_0}$ could be also described as $\mathbb{P}_\gamma \downharpoonright (\gamma \setminus \delta_0)$ with this notation.

The poset $\mathbb{P}_\gamma \,|\, \mathbb{G}_{\delta_0}$ is forcing equivalent to $\mathbb{P}_\gamma / \mathbb{G}_{\delta_0}$.

Lemma 5.9. *The mapping*

$$(5.44) \quad i_\downharpoonright : \mathbb{P}_\gamma / \mathbb{G}_{\delta_0} \to \mathbb{P}_\gamma \,|\, \mathbb{G}_{\delta_0} \;; \quad q \mapsto q \downharpoonright (\gamma \setminus \delta_0)$$

is a dense embedding.

Proof. $i_\downharpoonright$ is surjective: If $p \in \mathbb{P}_\gamma \,|\, \mathbb{G}_{\delta_0}$ then $p \in \mathbb{P}_\gamma / \mathbb{G}_{\delta_0}$ and $i_\downharpoonright(p) = p$.

$i_\downharpoonright(\mathbb{1}_{\mathbb{P}_\gamma / \mathbb{G}_{\delta_0}}) = i_\downharpoonright(\mathbb{1}_{\mathbb{P}_\gamma}) = \mathbb{1}_{\mathbb{P}_\gamma} \downharpoonright (\gamma \setminus \delta_0) = \mathbb{1}_{\mathbb{P}_\gamma} = \mathbb{1}_{\mathbb{P}_\gamma \,|\, \mathbb{G}_{\delta_0}}$.

$i_\downharpoonright$ is order preserving: Suppose that $q_0, q_1 \in \mathbb{P}_\gamma / \mathbb{G}_{\delta_0}$ and $q_0 \leq_{\mathbb{P}_\gamma} q_1$. Then $q_0 \restriction \delta_0, q_1 \restriction \delta_0 \in \mathbb{G}_{\delta_0}$ and $q_0 \restriction \delta_0 \leq_{\mathbb{P}_{\delta_0}} q_1 \restriction \delta_0$. It follows that

$$(5.45) \quad q_0 \restriction \delta_0 \wedge (\underbrace{q_0 \downharpoonright (\gamma \setminus \delta_0)}_{= i_\downharpoonright(q_0)}) \leq_{\mathbb{P}_\gamma} q_0 \restriction \delta_0 \wedge (\underbrace{q_1 \downharpoonright (\gamma \setminus \delta_0)}_{= i_\downharpoonright(q_1)}).$$

Thus, $i_\downharpoonright(q_0) \leq_{\mathbb{P}_\gamma \,|\, \mathbb{G}_{\delta_0}} i_\downharpoonright(q_1)$.

$i_{\downharpoonright}$ is incompatibility preserving: Suppose that $\mathbb{q}_0$, $\mathbb{q}_1 \in \mathbb{P}_\gamma/\mathbb{G}_{\delta_0}$ and, $i_{\downharpoonright}(\mathbb{q}_0)$ and $i_{\downharpoonright}(\mathbb{q}_1)$ are compatible in $\mathbb{P}_\gamma \,|\, \mathbb{G}_{\delta_0}$. Then, there is $\mathbb{r} \in \mathbb{P}_\gamma \,|\, \mathbb{G}_{\delta_0}$ such that $\mathbb{r} \leq_{\mathbb{P}_\gamma \,|\, \mathbb{G}_{\delta_0}} i_{\downharpoonright}(\mathbb{q}_0),\ i_{\downharpoonright}(\mathbb{q}_1)$. By the definition of $\leq_{\mathbb{P}_\gamma \,|\, \mathbb{G}_{\delta_0}}$, this means that there are $\mathbb{s}_0$, $\mathbb{s}_1 \in \mathbb{G}_{\delta_0}$ such that $\mathbb{s}_0 \wedge \mathbb{r} \leq_{\mathbb{P}_\gamma} \mathbb{s}_0 \wedge i_{\downharpoonright}(\mathbb{q}_0)$ and $\mathbb{s}_1 \wedge \mathbb{r} \leq_{\mathbb{P}_\gamma} \mathbb{s}_1 \wedge i_{\downharpoonright}(\mathbb{q}_1)$.

Let $\mathbb{s}_2 \in \mathbb{G}_{\delta_0}$ be such that $\mathbb{s}_2 \leq_{\mathbb{P}_{\delta_0}} \mathbb{s}_0,\ \mathbb{s}_1,\ \mathbb{q}_0 \restriction \delta_0,\ \mathbb{q}_1 \restriction \delta_0$. Then $\mathbb{s}_2 \wedge \mathbb{r} \in \mathbb{P}_\gamma/\mathbb{G}_{\delta_0}$ and $\mathbb{s}_2 \wedge \mathbb{r} \leq_{\mathbb{P}_\gamma} \mathbb{q}_0, \mathbb{q}_1$ by Lemma 5.8, (1). □ (Lemma 5.9)

Working further in $\mathsf{V}[\mathbb{G}_{\delta_0}]$, let

$$(5.46)\quad \mathbb{S}_{\delta_0,\gamma} = (\mathbb{P}_\gamma \,|\, \mathbb{G}_{\delta_0}) \downharpoonright S = \{\mathbb{p} \in \mathbb{P}_\gamma \,:\, \mathrm{supp}(\mathbb{p}) \subseteq S \setminus \delta_0\}$$

be the poset with the pre-ordering $\leq_{\mathbb{P}_\gamma \,|\, \mathbb{G}_{\delta_0}}$ restricted to it and with the designated maximal element $\mathbb{1}_{\mathbb{S}_{\delta_0,\gamma}} = \mathbb{1}_{\mathbb{P}_\gamma}$. Let

$$(5.47)\quad \mathbb{T}_{\delta_0,\gamma} = (\mathbb{P}_\gamma \,|\, \mathbb{G}_{\delta_0}) \downharpoonright T = \{\mathbb{p} \in \mathbb{P}_\gamma \,:\, \mathrm{supp}(\mathbb{p}) \subseteq T \setminus \delta_0\}$$

be the poset with the pre-ordering $\leq_{\mathbb{P}_\gamma \,|\, \mathbb{G}_{\delta_0}}$ restricted to it and with the designated maximal element $\mathbb{1}_{\mathbb{T}_{\delta_0,\gamma}} = \mathbb{1}_{\mathbb{P}_\gamma}$.

Lemma 5.10. *In* $\mathsf{V}[\mathbb{G}_{\delta_0}]$, *the mapping*

$$(5.48)\quad \pi_{\delta_0,\gamma} : \mathbb{S}_{\delta_0,\gamma} \times \mathbb{T}_{\delta_0,\gamma} \to \mathbb{P}_\gamma \,|\, \mathbb{G}_{\delta_0}\,;\quad \langle \mathbb{s}, \mathbb{t} \rangle \mapsto \mathbb{s} \wedge \mathbb{t}$$

is a projection.

Proof. $\pi_{\delta_0,\gamma} \models (5.32)$ is clear by the definition of $\pi_{\delta_0,\gamma}$.

To show that $\pi_{\delta_0,\gamma}$ is order-preserving, suppose that $\mathbb{s}' \leq_{\mathbb{P}_\gamma \,|\, \mathbb{G}_{\delta_0}} \mathbb{s}$ and $\mathbb{t}' \leq_{\mathbb{P}_\gamma \,|\, \mathbb{G}_{\delta_0}} \mathbb{t}$. Then, there are $\mathbb{u}_0$, $\mathbb{u}_1 \in \mathbb{G}_{\delta_0}$ such that $\mathbb{u}_0 \wedge \mathbb{s}' \leq_{\mathbb{P}_\gamma} \mathbb{u}_0 \wedge \mathbb{s}$ and $\mathbb{u}_1 \wedge \mathbb{t}' \leq_{\mathbb{P}_\gamma} \mathbb{u}_1 \wedge \mathbb{t}$.

Let $\mathbb{u}_2 \in \mathbb{G}_{\delta_0}$ be such that $\mathbb{u}_2 \leq_{\mathbb{P}_{\delta_0}} \mathbb{u}_0,\ \mathbb{u}_1$. By Lemma 5.8, (1), we have $\mathbb{u}_2 \wedge \mathbb{s}' \leq_{\mathbb{P}_\gamma} \mathbb{u}_2 \wedge \mathbb{s}$ and $\mathbb{u}_2 \wedge \mathbb{t}' \leq_{\mathbb{P}_\gamma} \mathbb{u}_2 \wedge \mathbb{t}$.

By Lemma 5.3, it follows that $\mathbb{u}_2 \wedge (\mathbb{s}' \wedge \mathbb{t}') \leq_{\mathbb{P}_\gamma} \mathbb{u}_2 \wedge (\mathbb{s} \wedge \mathbb{t})$. Thus, $\pi_{\delta_0,\gamma}(\langle \mathbb{s}', \mathbb{t}' \rangle) = \mathbb{s}' \wedge \mathbb{t}' \leq_{\mathbb{P}_\gamma \,|\, \mathbb{G}_{\delta_0}} \mathbb{s} \wedge \mathbb{t} = \pi_{\delta_0,\gamma}(\langle \mathbb{s}, \mathbb{t} \rangle)$.

To show that $\pi_{\delta_0,\gamma}$ also satisfies (5.34), suppose that $\langle \mathbb{s}, \mathbb{t} \rangle \in \mathbb{S}_{\delta_0,\gamma} \times \mathbb{T}_{\delta_0,\gamma}$ and $\mathbb{p} \in \mathbb{P}_\gamma \,|\, \mathbb{G}_{\delta_0}$ are such that

$$(5.49)\quad \mathbb{p} \leq_{\mathbb{P}_\gamma \,|\, \mathbb{G}_{\delta_0}} \mathbb{s} \wedge \mathbb{t} = \tau_{\delta_0,\gamma}(\langle \mathbb{s}, \mathbb{t} \rangle).$$

Let $\mathbb{u} \in \mathbb{G}_{\delta_0}$ be such that $\mathbb{u} \wedge \mathbb{p} \leq_{\mathbb{P}_\gamma} \mathbb{u} \wedge (\mathbb{s} \wedge \mathbb{t})$.

Let $\mathbb{p}_0$ be a $\gamma \setminus \delta_0$-sequence defined by

$$(5.50)\quad \mathbb{p}_0(\xi) = \begin{cases} \underset{\sim}{\mathbb{q}}_\xi, & \text{if } \xi \in \mathrm{supp}(\mathbb{p}); \\ \mathbb{1}_{\underset{\sim}{\mathbb{Q}}_\xi}, & \text{otherwise} \end{cases}$$

for $\xi \in \gamma \setminus \delta_0$, where $\underset{\sim}{\mathbb{q}}_\xi$ is a canonical $\mathbb{P}_\xi$-name of an element of $\underset{\sim}{\mathbb{Q}}_\xi$ such that

(5.51) $\quad \mathbb{u} \wedge \mathbb{p} \restriction \xi \Vdash_{\mathbb{P}_\xi}$ " $\underset{\sim}{\mathbb{q}}_\xi \equiv \mathbb{p}(\xi)$ ", and

(5.52) $\quad \mathbb{p}' \Vdash_{\mathbb{P}_\xi}$ " $\underset{\sim}{\mathbb{q}}_\xi \equiv (\mathbb{s} \wedge \mathbb{t})(\xi)$ ", $\qquad$ for all $\mathbb{p}' \in \mathbb{P}_\xi$ with $\mathbb{p}' \perp_{\mathbb{P}_\xi} \mathbb{u} \wedge \mathbb{p} \restriction \xi$.

Note that, by (5.51) and (5.52), we have

(5.53) $\quad \Vdash_{\mathbb{P}_\xi}$ " $\underset{\sim}{\mathbb{q}}_\xi \leq_{\underset{\sim}{\mathbb{Q}}_\xi} (\mathbb{s} \wedge \mathbb{t})(\xi)$ " for all $\xi \in \gamma \setminus \delta_0$.

Let $\mathbb{s}_0 = \mathbb{p}_0 \downharpoonright S$ and $\mathbb{t}_0 = \mathbb{p}_0 \downharpoonright T$.

By the construction, it is clear that the following claim holds, and this shows that $\langle \mathbb{s}_0, \mathbb{t}_0 \rangle$ is a witness for (5.34).

Claim 5.10.1. (a) $\pi_{\delta_0,\gamma}(\langle \mathbb{s}_0, \mathbb{t}_0 \rangle) = \mathbb{p}_0 \leq_{\mathbb{P}_\gamma \,|\, \mathbb{G}_{\delta_0}} \mathbb{p}$,
(b) $\langle \mathbb{s}_0, \mathbb{t}_0 \rangle \leq_{\mathbb{S}_{\delta_0,\gamma} \times \mathbb{T}_{\delta_0,\gamma}} \langle \mathbb{s}, \mathbb{t} \rangle$.

$\vdash$ (a): By (5.51). (b): By (5.53). $\dashv$ (Claim 5.10.1)

□ (Lemma 5.10)

Let $\underset{\sim}{\mathbb{G}}_{\delta_0}$ denote the standard $\mathbb{P}_{\delta_0}$-name of a $(\mathsf{V}, \mathbb{P}_{\delta_0})$-generic filter.

Lemma 5.11. *Suppose that $\underset{\sim}{\mathbb{q}}_0$ and $\underset{\sim}{\mathbb{q}}_1$ are $\mathbb{P}_{\delta_0}$-names of elements of $\mathbb{P}_\gamma \,|\, \underset{\sim}{\mathbb{G}}_{\delta_0}$. Then, we have*

(5.54) $\quad \Vdash_{\mathbb{P}_{\delta_0}}$ " $\underset{\sim}{\mathbb{q}}_0 \leq_{\mathbb{P}_\gamma \,|\, \underset{\sim}{\mathbb{G}}_{\delta_0}} \underset{\sim}{\mathbb{q}}_1$ " $\Leftrightarrow \Vdash_{\mathbb{P}_{\delta_0}}$ " $\vec{\mathbb{1}}_{\delta_0} \wedge \underset{\sim}{\mathbb{q}}_0 \leq_{\mathbb{P}_\gamma} \vec{\mathbb{1}}_{\delta_0} \wedge \underset{\sim}{\mathbb{q}}_1$ ".

Proof. "$\Leftarrow$" is trivial since $\Vdash_{\mathbb{P}_\delta}$ " $\vec{\mathbb{1}}_{\delta_0} \in \underset{\sim}{\mathbb{G}}_{\delta_0}$ ".

"$\Rightarrow$": Suppose that the left side of (5.54) holds. This means that $\Vdash_{\mathbb{P}_{\delta_0}}$ " $\exists p \in \underset{\sim}{\mathbb{G}}_{\delta_0} (p \wedge \underset{\sim}{\mathbb{q}}_0 \leq_{\mathbb{P}_\gamma} p \wedge \underset{\sim}{\mathbb{q}}_1)$ ". By Lemma 5.8, it follows that

(5.55) $\quad \Vdash_{\mathbb{P}_{\delta_0}}$ " $\exists p \in \underset{\sim}{\mathbb{G}}_{\delta_0} \forall p' \leq_{\mathbb{P}_\gamma} p \, (p' \wedge \underset{\sim}{\mathbb{q}}_0 \leq_{\mathbb{P}_\gamma} p' \wedge \underset{\sim}{\mathbb{q}}_1)$ ".

Suppose, toward a contradiction, that

(5.56) $\quad \not\Vdash_{\mathbb{P}_{\delta_0}}$ " $\vec{\mathbb{1}}_{\delta_0} \wedge \underset{\sim}{\mathbb{q}}_0 \leq_{\mathbb{P}_\gamma} \vec{\mathbb{1}}_{\delta_0} \wedge \underset{\sim}{\mathbb{q}}_1$ ".

Then, there are $\mathbb{p}_0 \in \mathbb{P}_{\delta_0}$ and $\delta_0 \leq \delta < \gamma$ such that, for any $\mathbb{p} \leq_{\mathbb{P}_{\delta_0}} \mathbb{p}_0$,

(5.57) $\quad \mathbb{p} \Vdash_{\mathbb{P}_{\delta_0}}$ " $\vec{\mathbb{1}}_{\delta_0} \wedge \underset{\sim}{\mathbb{q}}_0 \restriction \delta \leq_{\mathbb{P}_\delta} \vec{\mathbb{1}}_{\delta_0} \wedge \underset{\sim}{\mathbb{q}}_1 \restriction \delta$ ", but

(5.58) $\quad \mathbb{p} \Vdash_{\mathbb{P}_{\delta_0}} “\vec{\mathbb{1}}_{\delta_0} \wedge \underset{\sim}{\mathbb{q}}_0 \restriction \delta \Vdash_{\mathbb{P}_\delta} “\underset{\sim}{\mathbb{q}}_0(\delta) \not\leq_{\underset{\sim}{\mathbb{Q}}_\delta} \underset{\sim}{\mathbb{q}}_1(\delta)”\,”.$

By Lemma 5.8, (1), it follows that, for any $\mathbb{p} \leq_{\mathbb{P}_{\delta_0}} \mathbb{p}_0$, $\mathbb{p} \Vdash_{\mathbb{P}_{\delta_0}} “\check{\mathbb{p}} \,\varepsilon\, \underset{\sim}{\mathbb{G}}_{\delta_0}$ and $\check{\mathbb{p}} \wedge \underset{\sim}{\mathbb{q}}_0 \not\leq_{\mathbb{P}_\gamma} \check{\mathbb{p}} \wedge \underset{\sim}{\mathbb{q}}_1”$. This is a contradiction to (5.55) by Lemma 5.8, (1). □ (Lemma 5.11)

Lemma 5.12. *For* $\delta_0 < \gamma \leq \kappa$ *and* $(\mathsf{V}, \mathbb{P}_{\delta_0})$*-generic filter* $\mathbb{G}_{\delta_0}$*, we have* $\mathsf{V}[\mathbb{G}_{\delta_0}] \models$ *“*$\mathbb{S}_{\delta_0,\gamma}$ *is* $<\nu(\delta_0)$*-closed”.*

Proof. In V, let $\underset{\sim}{\mathbb{S}}_{\delta_0,\gamma}$ be a $\mathbb{P}_{\delta_0}$-name of $\mathbb{S}_{\delta_0,\gamma}$ and $\underset{\sim}{h}$ be a $\mathbb{P}_{\delta_0}$-name of a descending δ-sequence in $\mathbb{S}_{\delta_0,\gamma}$ for some

(5.59) $\quad \delta < \nu(\delta_0).$

By Lemma 5.11, we have $\Vdash_{\mathbb{P}_{\delta_0}} “\vec{\mathbb{1}}_{\delta_0} \wedge \underset{\sim}{h}(\xi) \leq_{\mathbb{P}_\gamma} \vec{\mathbb{1}}_{\delta_0} \wedge \underset{\sim}{h}(\eta)”$ for all $\xi < \eta < \delta$.

Let

(5.60) $\quad D = \{\alpha < \gamma : \mathbb{r} \Vdash_{\mathbb{P}_{\delta_0}} “\alpha \in \mathrm{supp}(\underset{\sim}{h}(\xi))$ for some $\xi < \delta”$ for some $\mathbb{r} \in \mathbb{P}_{\delta_0}\}$.

Since $\Vdash_{\mathbb{P}_{\delta_0}} “(\forall \xi < \delta)\, \mathrm{supp}(\xi) \subseteq S \setminus \nu(\delta_0)”$, we have

(5.61) $\quad D \subseteq S \setminus \nu(\delta_0).$

Claim 5.12.1. *For any regular* $\delta_0 \leq \mu \leq \gamma$*, we have* $|\, D \cap \mu \,| < \mu$*. Thus,* $D \in \mathcal{I}_\gamma$*.*

$\vdash$ By (5.61), it is enough to show the inequality for all regular cardinal μ with $\nu(\delta_0) \leq \mu \leq \gamma$. For such μ, we have $\Vdash_{\mathbb{P}_{\delta_0}} “\,\mathrm{supp}(\underset{\sim}{h}(\xi)) \cap \mu$ is a bounded subset of μ” for all $\xi < \delta$. Thus

(5.62) $\quad D_{\mu,\xi,\mathbb{r}} = \{\alpha < \mu : \mathbb{r} \Vdash_{\mathbb{P}_{\delta_0}} “\alpha \in \mathrm{supp}(\underset{\sim}{h}(\xi))”\}$

is a bounded subset of μ for each $\xi < \delta$ and $\mathbb{r} \in \mathbb{P}_{\delta_0}$. By (5.59) and Lemma 5.4 for $\nu = \nu(\alpha)$, it follows that $D \cap \mu = \bigcup_{\xi<\delta, \mathbb{r}\in\mathbb{P}_{\delta_0}} D_{\mu,\xi,\mathbb{r}}$ is a bounded subset of μ. $\dashv$ (Claim 5.12.1)

Now we define, by induction on $\delta_0 \leq i \leq \gamma$, $\mathbb{P}_{\delta_0}$-names $\underset{\sim}{\mathbb{p}}_i$, $i \in \gamma+1 \setminus \delta_0$ such that

(5.63) $\quad \Vdash_{\mathbb{P}_{\delta_0}} “\underset{\sim}{\mathbb{p}}_i \,\varepsilon\, \mathbb{P}_i \downharpoonright (S \setminus \delta_0)”$ for all $i \in \gamma + 1 \setminus \delta_0$;

(5.64) $\quad \Vdash_{\mathbb{P}_{\delta_0}} “\,\mathrm{supp}(\underset{\sim}{\mathbb{p}}_i) \subseteq \check{D}”$ for all $i \in \gamma + 1 \setminus \delta_0$;

(5.65) $\Vdash_{\mathbb{P}_{\delta_0}}$ "$(\langle \underset{\sim}{p}_i : i \in \gamma+1 \setminus \delta_0\rangle)^\bullet$ is an increasing sequence of sequences" ;

and

(5.66) $\Vdash_{\mathbb{P}_{\delta_0}}$ "$\underset{\sim}{p}_i$ is a lower bound of $\langle \underset{\sim}{h}(\xi) \restriction i : \xi < \delta\rangle$ with respect to $\leq_{\mathbb{P}_i \mid \underset{\sim}{\mathbb{G}}_{\delta_0}}$"

for all $i \in \gamma+1 \setminus \delta_0$.

For $i = \delta_0$, $p_i = \emptyset$ will do.

Suppose now that i is a limit ordinal and $\underset{\sim}{p}_j$, $j < i$ has been defined such that

(5.63′) $\Vdash_{\mathbb{P}_{\delta_0}}$ "$\underset{\sim}{p}_j \,\varepsilon\, \mathbb{P}_j \downharpoonright (S \setminus \delta_0)$" for all $j \in i \setminus \delta_0$;

(5.64′) $\Vdash_{\mathbb{P}_{\delta_0}}$ " $\mathrm{supp}(\underset{\sim}{p}_j) \subseteq \check{D}$" for all $j \in i \setminus \delta_0$;

(5.65′) $\Vdash_{\mathbb{P}_{\delta_0}}$ "$(\langle \underset{\sim}{p}_j : j \in i \setminus \delta_0\rangle)^\bullet$ is an increasing sequence of sequences" ;

and

(5.66′) $\Vdash_{\mathbb{P}_{\delta_0}}$ "$\underset{\sim}{p}_j$ is a lower bound of $\langle \underset{\sim}{h}(\xi) \restriction j : \xi < \delta\rangle$ with respect to $\leq_{\mathbb{P}_j \mid \underset{\sim}{\mathbb{G}}_{\delta_0}}$"

for all $j \in i \setminus \delta_0$.

By Lemma 5.11, (5.66′) implies

(5.67) $\Vdash_{\mathbb{P}_{\delta_0}}$ "$\vec{\mathbb{1}}_{\delta_0} \wedge \underset{\sim}{p}_j \leq_{\mathbb{P}_j} \vec{\mathbb{1}}_{\delta_0} \wedge \underset{\sim}{h}(\xi) \restriction j$" for all $\xi < \delta$ and $j < i$.

Let $\underset{\sim}{p}_i$ be the $\mathbb{P}_{\delta_0}$-name such that

(5.68) $\Vdash_{\mathbb{P}_{\delta_0}}$ "$\underset{\sim}{p}_i \equiv \bigcup(\{\underset{\sim}{p}_j : j < i\})^\bullet$".

We show that $\underset{\sim}{p}_i$ together with $\underset{\sim}{p}_j$, $j < i$ satisfies

(5.63″) $\Vdash_{\mathbb{P}_{\delta_0}}$ "$\underset{\sim}{p}_j \,\varepsilon\, \mathbb{P}_j \downharpoonright (S \setminus \delta_0)$" for all $j \in i+1 \setminus \delta_0$;

(5.64″) $\Vdash_{\mathbb{P}_{\delta_0}}$ " $\mathrm{supp}(\underset{\sim}{p}_j) \subseteq \check{D}$" for all $j \in i+1 \setminus \delta_0$;

(5.65″) $\Vdash_{\mathbb{P}_{\delta_0}}$ "$(\langle \underset{\sim}{p}_j : j \in i+1 \setminus \delta_0\rangle)^\bullet$ is an increasing sequence of sequences" ;

and

(5.66″) $\Vdash_{\mathbb{P}_{\delta_0}}$ "$\underset{\sim}{p}_j$ is a lower bound of $\langle \underset{\sim}{h}(\xi) \restriction j : \xi < \delta\rangle$ with respect to $\leq_{\mathbb{P}_j \mid \underset{\sim}{\mathbb{G}}_{\delta_0}}$"

for all $j \in i+1 \setminus \delta_0$.

(5.64″) follows from (5.64′) and (5.68). (5.63″) follows from this. (5.65″) is clear by (5.68) and (5.66″) follows from (5.67).

Finally, suppose that $\underset{\sim}{p}_j$, $j \leq i$ has been defined for $\nu(\delta_0) \leq j < \gamma$ in accordance with (5.63) $\sim$ (5.66). In particular, we have

$$(5.67') \quad \Vdash_{\mathbb{P}_{\delta_0}} \text{“}\, \vec{\mathbb{1}}_{\delta_0} \wedge \underset{\sim}{p}_i \leq_{\mathbb{P}_i} \vec{\mathbb{1}}_{\delta_0} \wedge \underset{\sim}{h}(\xi) \restriction i \,\text{” for all } \xi < \delta.$$

If $i \notin S$, then let $\underset{\sim}{p}_{i+1} = (\underset{\sim}{p}_{i+1} \cup \{\langle i, \mathbb{1}_{\underset{\sim}{\mathbb{Q}}_i}\rangle\})^{\bullet}_{\mathbb{P}_{\delta_0}}$.

If $i \in S$, then we have

$$(5.69) \quad \Vdash_{\mathbb{P}_i} \text{“}\, \underset{\sim}{\mathbb{Q}}_i \text{ is } {<}\,\nu(\delta_0)\text{-closed}\,\text{”}$$

by (B), (a) and (b) in the definition of our mixed support iteration. By (5.67′) and by the choice of $\underset{\sim}{h}$, we have

$$(5.70) \quad \Vdash_{\mathbb{P}_{\delta_0}} \text{“}\, \vec{\mathbb{1}}_{\delta_0} \wedge \underset{\sim}{p}_i \Vdash_{\mathbb{P}_i} \text{“}\, (\langle \underset{\sim}{h}(\xi)(i) \,:\, \xi < \delta\rangle)^{\bullet}_{\mathbb{P}_i} \text{ is a descending sequence in } \underset{\sim}{\mathbb{Q}}_i \,\text{” ”}.$$

By (5.69), there is a $\mathbb{P}_{\delta_0}$-name $\underset{\sim}{q}$ of $\mathbb{P}_{\delta_0}$-name such that

$$(5.71) \quad \Vdash_{\mathbb{P}_{\delta_0}} \text{“}\, \vec{\mathbb{1}}_{\delta_0} \wedge \underset{\sim}{p}_i \Vdash_{\mathbb{P}_i} \text{“}\, \underset{\sim}{q} \text{ is a lower bound of } (\langle \underset{\sim}{h}(\xi)(i) \,:\, \xi < \delta\rangle)^{\bullet}_{\mathbb{P}_i} \text{” ”}.$$

Let $\underset{\sim}{p}_{i+1} = (\underset{\sim}{p}_i \cup \{\langle i, \underset{\sim}{q}\rangle\})^{\bullet}_{\mathbb{P}_{\delta_0}}$. Similarly to the previous case, we can show that $\underset{\sim}{p}_{i+1}$ together with $\underset{\sim}{p}_j$, $j \leq i$ satisfies (5.63) $\sim$ (5.66).

□ (Lemma 5.12)

Lemma 5.13. *Suppose that $\mathbb{P}$ is a poset, $\underset{\sim}{\mathbb{Q}}$ a $\mathbb{P}$-name of a poset with*

$$(5.72) \quad \Vdash_{\mathbb{P}} \text{“}\, \underset{\sim}{\mathbb{Q}} \text{ is ccc}\,\text{”},$$

and $\mathbb{S}$ is a σ-closed poset. Then we have

$$(5.73) \quad \Vdash_{\mathbb{S}} \text{“}\, \Vdash_{\check{\mathbb{P}}} \text{“}\, \check{\underset{\sim}{\mathbb{Q}}} \text{ is ccc}\,\text{” ”}.$$

Proof. Suppose that $\underset{\sim}{S}$ is a $\mathbb{S}$-name of a $\mathbb{P}$-name such that

$$(5.74) \quad \Vdash_{\mathbb{S}} \text{“}\, \Vdash_{\check{\mathbb{P}}} \text{“}\, \underset{\sim}{S} \text{ is a subset of } \check{\underset{\sim}{\mathbb{Q}}} \text{ of cardinality } \aleph_1 \,\text{” ”}.$$

We have to show

(5.75) $\Vdash_{\mathbb{S}}$ " $\Vdash_{\check{\mathbb{P}}}$ " there are compatible elements in $\underset{\sim}{S}$ " ".

Let $\underset{\sim}{f}$ be a $\mathbb{S}$-name of $\mathbb{P}$-name such that

(5.76) $\Vdash_{\mathbb{S}}$ " $\Vdash_{\check{\mathbb{P}}}$ " $\underset{\sim}{f} : \omega_1 \to \underset{\sim}{S}$ and
$\underset{\sim}{f}$ is an injective enumeration of $\underset{\sim}{S}$ " ".

Let $\mathbb{s} \in \mathbb{S}$ and $\mathbb{p} \in \mathbb{P}$ be arbitrary. By σ-closedness of $\mathbb{S}$, we can find a decreasing sequence $\langle \mathbb{s}_\alpha : \alpha < \omega_1 \rangle$ of elements of $\mathbb{S}$ and a sequence $\underset{\sim}{\mathbb{q}}_\alpha$, $\alpha < \omega_1$ of $\mathbb{P}$-names such that

(5.77) $\mathbb{s}_0 \leq_{\mathbb{S}} \mathbb{s}$,
(5.78) $\mathbb{s}_\alpha \Vdash_{\mathbb{S}}$ " $\Vdash_{\check{\mathbb{P}}}$ " $\check{\underset{\sim}{\mathbb{q}}}_\alpha \equiv \underset{\sim}{f}(\alpha)$ " ".

By (5.74), (5.76) and (5.78), we have

(5.79) $\mathbb{s}_\alpha \Vdash_{\mathbb{S}}$ " $\Vdash_{\check{\mathbb{P}}}$ " $\check{\underset{\sim}{\mathbb{q}}}_\alpha \,\varepsilon\, \check{\underset{\sim}{\mathbb{Q}}}$ " ".

Since the relation $\cdot \Vdash_{\cdot}$ "$\cdot\, \varepsilon \,\cdot$" is Δ_1, it follows that

(5.80) $\Vdash_{\mathbb{P}}$ " $\underset{\sim}{\mathbb{q}}_\alpha \,\varepsilon\, \underset{\sim}{\mathbb{Q}}$ ".

By (5.72), there are $\mathbb{p}' \leq_{\mathbb{P}} \mathbb{p}$ and $\alpha_0 < \alpha_1 < \omega_1$ such that

(5.81) $\mathbb{p}' \Vdash_{\mathbb{P}}$ " $\underset{\sim}{\mathbb{q}}_{\alpha_0} \top_{\underset{\sim}{\mathbb{Q}}} \underset{\sim}{\mathbb{q}}_{\alpha_1}$ ".

By (5.76) and (5.78), and since $\langle \mathbb{s}_\alpha : \alpha < \omega_1 \rangle$ is decreasing,

(5.82) $\mathbb{s}_{\alpha_1} \Vdash_{\mathbb{S}}$ " $\mathbb{p}' \Vdash_{\mathbb{P}}$ " $\underset{\sim}{f}(\alpha_0) \top_{\underset{\sim}{\mathbb{Q}}} \underset{\sim}{f}(\alpha_1)$ " ".

Thus

(5.83) $\mathbb{s}_{\alpha_1} \Vdash_{\mathbb{S}}$ " $\exists x \leq_{\check{\mathbb{P}}} \mathbb{p}\; x \Vdash_{\check{\mathbb{P}}}$ " there are compatible elements in $\underset{\sim}{S}$ " ".

$\mathbb{s}_{\alpha_1} \leq_{\mathbb{S}} \mathbb{s}$ by (5.77). Since $\mathbb{s}$ was arbitrary, if follows that

(5.84) $\Vdash_{\mathbb{S}}$ " $\exists x \leq_{\check{\mathbb{P}}} \check{\mathbb{p}}\; x \Vdash_{\check{\mathbb{P}}}$ " there are compatible elements in $\underset{\sim}{S}$ " ".

Now, since $\mathbb{p}$ was arbitrary, (5.75) follows. □ (Lemma 5.13)

Lemma 5.14. (1) *For $\delta \leq \kappa$, ι_δ is an isomorphism from $\mathbb{O}_\delta$ to $\mathbb{T}_{0,\delta}$. $\langle i_\beta : \beta \leq \delta \rangle$ forms a commutative system together with $\langle \mathbb{O}_\beta, i^*_{\beta,\gamma} : \beta \leq \gamma \leq \delta \rangle$ and $\langle \mathbb{T}_{0,\beta}, i_{\beta,\gamma} \restriction \mathbb{T}_\beta : \beta \leq \gamma \leq \delta \rangle$. In particular, $\langle \mathbb{T}_{0,\beta} : \beta \leq \delta \rangle$ is* *homomorphic to the sequence of iterands of a finite support iteration of ccc posets.*

(2) *For* $\delta_0 < \gamma \leq \kappa$ *and* $(\mathsf{V}, \mathbb{P}_{\delta_0})$*-generic filter* $\mathbb{G}_{\delta_0}$*, we have*

(5.85) $\mathsf{V}[\mathbb{G}_{\delta_0}] \models$ "$\mathbb{T}_{\delta_0,\gamma}$ *has the ccc*".

Proof. (1): By induction $\delta \leq \kappa$.

(2): Note first that, by Lemma 5.6, $\pi_{0,\delta_0} : \mathbb{S}_{0,\delta_0} \times \mathbb{T}_{0,\delta_0} \to \mathbb{P}_{\delta_0}$ is a projection. Let $\mathbb{G}_{\mathbb{S}} * \mathbb{G}_{\mathbb{T}}$ be a $(\mathsf{V}, \mathbb{S}_{0,\delta_0} \times \mathbb{T}_{0,\delta_0}/\mathbb{G}_{\delta_0})$-generic filter in the sense of Lemma 5.5, (2) where we assume that $\mathbb{G}_{\mathbb{S}}$ and $\mathbb{G}_{\mathbb{T}}$ are the generic filters over $\mathbb{S}_{0,\delta_0}$ and $\mathbb{T}_{0,\delta_0}$ respectively. We have $\mathbb{G}_{\mathbb{T}} = \mathbb{G}_{\delta_0} \restriction T$. Thus $\mathbb{T}_{\delta_0,\gamma} = \mathbb{T}_{0,\gamma} | \mathbb{G}_{\mathbb{T}}$. By the Factor Lemma for finite support iteration of ccc posets, we have $\mathsf{V}[\mathbb{G}_{\mathbb{T}}] \models \mathbb{T}_{\delta_0,\gamma}$ is ccc.

Since $\mathbb{S}_{0,\delta_0}$ is σ-closed by Lemma 5.12, $\mathsf{V}[\mathbb{G}_{\mathbb{S}}][\mathbb{G}_{\mathbb{T}}] \models \mathbb{T}_{\delta_0,\gamma}$ is ccc by Lemma 5.13. Since $\mathsf{V}[\mathbb{G}_{\delta_0}]$ is an inner model of $\mathsf{V}[\mathbb{G}_{\mathbb{S}}][\mathbb{G}_{\mathbb{T}}]$, it follows that $\mathsf{V}[\mathbb{G}_{\delta_0}] \models \mathbb{T}_{\delta_0,\gamma}$ is ccc. ❑ (Lemma 5.14)

Summarizing what we have proved above, we obtain the following:

Proposition 5.15. *Suppose that* κ *is a supercompact cardinal,* $f : \kappa \to \kappa$ *a Laver function with* $S, T \subseteq \kappa$ *defined by* (5.3), (5.4), *and let* $\nu : \kappa \to \kappa$ *be defined by* (5.5).

For the preparatory finite support ccc iteration $\langle \mathbb{O}_\alpha, \underset{\sim}{\mathbb{R}}_\beta : \alpha \leq \kappa, \beta < \kappa \rangle$ *satisfying* (5.6), (5.7), *let* $\langle \mathbb{P}_\alpha, \underset{\sim}{\mathbb{Q}}_\beta : \alpha \leq \kappa, \beta < \kappa \rangle$ *be the Easton-type mixed support iteration over* $\langle \mathbb{O}_\alpha, \underset{\sim}{\mathbb{R}}_\beta : \alpha \leq \kappa, \beta < \kappa \rangle$ *as defined in* (A) *and* (B) *on pages 121* ~ *123 with the complete embeddings* $i_{\delta,\gamma}$*,* ι_δ *and projections* $p_{\delta,\gamma}$ *for* $\delta \leq \gamma \leq \kappa$.

For any $(\mathsf{V}, \mathbb{P}_\kappa)$*-generic filter* $\mathbb{G}_\kappa$*,* $\delta_0 < \kappa$ *and* $\mathbb{G}_{\delta_0} = p_{\kappa,\delta_0}{}''\mathbb{G}_\kappa$*, there are posets* $\mathbb{S}$*,* $\mathbb{T}$*, a regular subposet* $\mathbb{Q}$ *of the completion of* $\mathbb{S} \times \mathbb{T}$ *in* $\mathsf{V}[\mathbb{G}_{\delta_0}]$ *such that*

(5.86) $\mathsf{V}[\mathbb{G}_{\delta_0}] \models$ "$\mathbb{S}$ *is* $\nu(\delta_0)$*-closed and* $\mathbb{T}$ *is ccc*", *and*

(5.87) $\mathsf{V}[\mathbb{G}_{\delta_0}] \models$ "$\mathbb{Q} \sim \mathbb{P}_\kappa/\mathbb{G}_{\delta_0}$".

In particular, there is a $(\mathsf{V}[\mathbb{G}_{\delta_0}], RO(\mathbb{S} \times \mathbb{T}))$*-generic filter* $\tilde{\mathbb{H}}$ *such that, by letting* $\mathbb{H} = \tilde{\mathbb{H}} \cap \mathbb{Q}$*, we have* $\mathsf{V}[\mathbb{G}_\kappa] = \mathsf{V}[\mathbb{G}_{\delta_0}][\mathbb{H}]$.

Proof. By Lemma 5.9, we have $\mathsf{V}[\mathbb{G}_{\delta_0}] \models \mathbb{P}_\kappa/\mathbb{G}_{\delta_0} \sim \mathbb{P}_\kappa | \mathbb{G}_{\delta_0}$.

In $V[\mathbb{G}_{\delta_0}]$, $\mathbb{P}_\kappa | \mathbb{G}_{\delta_0}$ is forcing equivalent to a regular sub-poset of the completion of $\mathbb{S}_{\delta_0,\gamma} \times \mathbb{T}_{\delta_0,\gamma}$ by Lemma 5.10 (c.f. Lemma 5.7). By Lemma 5.12, $\mathbb{S}_{\delta_0,\gamma}$ is $\nu(\delta_0)$-closed and by Lemma 5.14, $\mathbb{T}_{\delta_0,\gamma}$ is ccc. ❑ (Proposition 5.15)

Theorem 5.16. *Suppose that* κ, f, S, T, ν, $\langle \mathbb{O}_\alpha, \underset{\sim}{\mathbb{R}}_\beta : \alpha \leq \kappa, \beta < \kappa \rangle$, $\langle \mathbb{P}_\alpha, \underset{\sim}{\mathbb{Q}}_\beta : \alpha \leq \kappa, \beta < \kappa \rangle$, $i_{\delta,\gamma}$, ι_δ, $p_{\delta,\gamma}$ *for* $\delta \leq \gamma \leq \kappa$, *and* $\mathbb{G}_\kappa$ *are as in* Proposition 5.15.

(0) *If* $\langle \mathbb{O}_\alpha, \underset{\sim}{\mathbb{R}}_\beta : \alpha \leq \kappa, \beta < \kappa \rangle$ *adds* κ *many reals then* $\mathsf{V}[\mathbb{G}_\kappa] \models \kappa = 2^{\aleph_0}$.

(1) *In* $\mathsf{V}[\mathbb{G}_\kappa]$, κ *is strongly Laver-generically supercompact for* $(\mathfrak{P}, \mathfrak{Q}_\theta)$ *for all* $\theta \in \mathsf{Card}$ *for the properties of posets* $\mathfrak{P}$ *and* $\mathfrak{Q}_\theta$ *as in* Proposition 4.7, (2).

(1′) *If* κ *is superhuge, then, in* $\mathsf{V}[\mathbb{G}_\kappa]$, κ *is strongly Laver-generically superhuge for* $(\mathfrak{P}, \mathfrak{Q}_\theta)$ *for all* $\theta \in \mathsf{Card}$ *for the properties of posets* $\mathfrak{P}$ *and* $\mathfrak{Q}_\theta$ *as in* Proposition 4.7, (2).

(2) *If the preparatory iteration* $\langle \mathbb{O}_\alpha, \underset{\sim}{\mathbb{R}}_\beta : \alpha \leq \kappa, \beta < \kappa \rangle$ *is such that* $\Vdash_{\mathbb{Q}_\alpha}$ "$\underset{\sim}{\mathbb{R}}_\alpha \equiv \mathrm{Fn}(\omega, 2)$ " *for all* $\alpha \in T$, *then in* $\mathsf{V}[\mathbb{G}_\kappa]$, κ *is strongly Laver-generically supercompact for* $(\mathfrak{P}, \mathfrak{Q}_\theta)$ *for all* $\theta \in \mathsf{Card}$ *for the properties of posets* $\mathfrak{P}$ *and* $\mathfrak{Q}_\theta$ *as in* Proposition 4.7, (1).

(2′) *If the preparatory iteration* $\langle \mathbb{O}_\alpha, \underset{\sim}{\mathbb{R}}_\beta : \alpha \leq \kappa, \beta < \kappa \rangle$ *is such that* $\Vdash_{\mathbb{Q}_\alpha}$ "$\underset{\sim}{\mathbb{R}}_\alpha \equiv \mathrm{Fn}(\omega, 2)$ " *for all* $\alpha \in T$ *and* κ *is superhuge, then in* $\mathsf{V}[\mathbb{G}_\kappa]$, κ *is strongly Laver-generically superhuge for* $(\mathfrak{P}, \mathfrak{Q}_\theta)$ *for all* $\theta \in \mathsf{Card}$ *for the properties of posets* $\mathfrak{P}$ *and* $\mathfrak{Q}_\theta$ *as in* Proposition 4.7, (1).

Proof. (0): By Lemma 5.4, κ is a regular cardinal in $\mathsf{V}[\mathbb{G}_\kappa]$ and $2^{\aleph_0} \leq \kappa$ in $\mathsf{V}[\mathbb{G}_\kappa]$. Since $\iota_\kappa : \mathbb{O}_\kappa \to \mathbb{P}_\kappa$ is a complete embedding, if $\mathbb{O}_\kappa$ adds κ many reals then $\kappa \geq 2^{\aleph_0}$.

(1): In $V[\mathbb{G}_\kappa]$, let $\mathbb{P} = \mathrm{Col}(\kappa, \mu)$ for some cardinal μ and, let λ and θ be two other cardinals. Without loss of generality, we may assume that $\mu^{<\kappa} \leq \lambda$. Since f is a Laver function there is an elementary embedding $j : \mathsf{V} \overset{\preccurlyeq}{\to} M$ such that

(5.88) $\quad crit(j) = \kappa$,

(5.89) $\quad j(\kappa) > \lambda$,

(5.90) $\quad [M]^\lambda \subseteq M$, and

(5.91) $\quad j(f)(\kappa) = \langle \mu, \theta, \emptyset \rangle$.

Let $S^* = j(S)$, $\nu^* = j(\nu)$ and, let $\vec{\mathbb{P}}^* = j(\langle \mathbb{P}_\alpha, \underset{\sim}{\mathbb{Q}}_\beta : \alpha \leq \kappa, \beta < \kappa \rangle)$. Since $\vec{\mathbb{P}}^*$ is $j(\kappa)$-(double) sequence by elementarity, we write

(5.92) $\quad \vec{\mathbb{P}}^* = \langle \mathbb{P}^*_\alpha, \underset{\sim}{\mathbb{Q}}^*_\beta : \alpha \leq j(\kappa), \beta < j(\kappa) \rangle$.

By the elementarity of j, Lemma 5.4, (5.88) and (5.90), we have $\mathbb{P}^*_\alpha = \mathbb{P}_\alpha$ and $\underset{\sim}{\mathbb{Q}}^*_\beta = \underset{\sim}{\mathbb{Q}}_\beta$ for all α, $\beta < \kappa$. $\mathbb{P}^*_\kappa = \mathbb{P}_\kappa$ by elementarity and (5.90). Note that $\kappa \in S^*$ by elementarity. So it follows that $\underset{\sim}{\mathbb{Q}}^*_\kappa = (\mathrm{Col}(\kappa,\mu))^\bullet_{\mathbb{P}_\kappa}$ by (5.22) and (5.91). Thus $\underset{\sim}{\mathbb{Q}}^*_\kappa[\mathbb{G}_\kappa] = \mathbb{P}$.

Also by (5.91), we have $\nu^*(\kappa) \geq \theta$. Let $\mathbb{g}$ be a $(\mathsf{V}[\mathbb{G}_\kappa], \mathbb{P})$-generic filter. In $M[\mathbb{G}_\kappa][\mathbb{g}]$, $\mathbb{P}^*_{j(\kappa)}/\mathbb{G}_\kappa * \mathbb{g}$ is forcing equivalent to a regular sub-poset of the completion of a poset of the form "ccc poset $\times$ θ-closed poset" by Proposition 5.15. Thus, $\mathbb{P}^*_{j(\kappa)}/\mathbb{G}_\kappa * \mathbb{g} \models \mathfrak{Q}_\theta$.

Let $\mathbb{H}^*$ be a $(\mathsf{V}[\mathbb{G}_\kappa * \mathbb{g}], \mathbb{P}^*_{j(\kappa)}/\mathbb{G}_\kappa * \mathbb{g})$-generic filter. Then we can find a $(\mathsf{V}, \mathbb{P}^*_{j(\kappa)})$-generic filter $\mathbb{H}$ such that $M[\mathbb{H}] = M[\mathbb{G}_\kappa * \mathbb{g} * \mathbb{H}^*]$ and $i^*_{\kappa,j(\kappa)}{}''\mathbb{G}_\kappa \subseteq \mathbb{H}$ for the complete embedding $i^*_{\kappa,j(\kappa)} : \mathbb{P}_\kappa \overset{\lessdot}{\to} \mathbb{P}^*_{j(\kappa)}$ associated with $\vec{\mathbb{P}}^*$. j can be then lifted to

$$(5.93) \quad \tilde{j} : \mathsf{V}[\mathbb{G}_\kappa] \overset{\preceq}{\to} M[\mathbb{H}];\ \underset{\sim}{a}[\mathbb{G}] \mapsto j(\underset{\sim}{a})[\mathbb{H}].$$

It is easy to show that $\tilde{j}$ with $\mathbb{H}$ satisfies (4.1) $\sim$ (4.5) and (4.6). The last condition holds by Lemma 4.1. This shows that κ is strongly Laver-generically supercompact for $(\mathfrak{P}, \mathfrak{Q}_\theta)$.

$(2')$: is proved similarly to (2) above. The condition $(4.6')$ is shown using Lemma 4.2.

(3), $(3')$: can be proved similarly to (2) and $(2')$. ☐ (Theorem 5.16)

6. Models with strong reflection properties down to $< 2^{\aleph_0}$ and with even stronger reflection properties but down to $\leq 2^{\aleph_0}$

As an application of the forcing constructions considered in the previous sections, we give two models of large continuum with strong reflection properties around the continuum. In one of the models, we have $\mathsf{HH}(<2^{\aleph_0})$ and, in the other, this reflection property is negated. Thus, we obtain the independence of $\mathsf{HH}(<2^{\aleph_0})$ from other strong reflection principles in the large continuum context.

In contrast, the most of the other reflection properties are situated in a tight web of implications which is (almost) upward directed (see e.g. the diagram in the last section of [Fuchino, Sakai and Ottenbreit[9]]). This suggests that the reflection of non-metrizability is a totally different kind of reflection statement from the other reflection principles.

With an arbitrary preparatory finite support ccc iteration $\langle \mathbb{O}_\alpha, \underset{\sim}{\mathbb{R}}_\beta :$

$\alpha \leq \kappa, \beta < \kappa\rangle$ we already have the following:

Theorem 6.1. *Let κ, κ_1 with $\kappa < \kappa_1$ be two supercompact cardinals and let $\langle \mathbb{P}_\alpha, \underset{\sim}{\mathbb{Q}}_\beta : \alpha \leq \kappa, \beta < \kappa\rangle$ be the Easton-type mixed support iteration over an arbitrary preparatory finite support ccc iterating $\langle \mathbb{O}_\alpha, \underset{\sim}{\mathbb{R}}_\beta : \alpha \leq \kappa, \beta < \kappa\rangle$ which adds κ many reals. Let $\mathbb{P} = \mathbb{P}_\kappa * (\mathrm{Col}(\kappa, \kappa_1))^{\bullet}_{\mathbb{P}_\kappa}$. Then, in the $\mathbb{P}$ generic extension over V, we have*

(6.1) $2^{\aleph_0} = \kappa$;

(6.2) $\mathsf{SDLS}^{int}_{+}(\mathcal{L}^{\aleph_0}_{stat}, < 2^{\aleph_0})$ *and* $\mathsf{GRP}^{<2^{\aleph_0}}(\leq 2^{\aleph_0})$.

(6.3) $\mathsf{SDLS}^{int}_{+}(\mathcal{L}^{\mathsf{PKL}}_{stat}, < 2^{\aleph_0})$ *and* $\mathcal{P}_{2^{\aleph_0}}(\lambda)$ *carries a σ-saturated normal ideal for all* $\lambda \geq 2^{\aleph_0}$.

Proof. (6.1): By Theorem 5.16, (0).

(6.2): By Theorem 5.16, (1), κ is strongly Laver-generically supercompact for $(\mathfrak{P}, \mathfrak{Q})$ for properties $\mathfrak{P}$, $\mathfrak{Q}$ as in Corollary 4.5. Thus, by Corollary 4.5, (2), we have $\mathsf{SDLS}^{int}_{+}(\mathcal{L}^{\aleph_0}_{stat}, < 2^{\aleph_0})$ and $\mathsf{GRP}^{<2^{\aleph_0}}(\leq 2^{\aleph_0})$ in the generic extension.

(6.3): By Theorem 5.16, (1), κ is strongly Laver-generically supercompact for $(\mathfrak{P}, \mathfrak{Q}_\theta)$ for any θ for $\mathfrak{P}$ and $\mathfrak{Q}_\theta$ as in Proposition 4.7, (2). Thus (6.3) holds by Proposition 4.7, (2) and (3). □ (Theorem 6.1)

The following strengthening of the forcing axiom $\mathsf{MA}(\mathcal{P})$ for a class $\mathcal{P}$ of posets was studied in [Fuchino, Ottenbreit and Sakai [10]]:

For a poset $\mathbb{P}$, $\mathbb{P}$-name $\underset{\sim}{S}$ of a set of subsets of On and a filter $\mathbb{G}$ on $\mathbb{P}$, let

$$(6.4)\quad \underset{\sim}{S}(\mathbb{G}) = \{b : b = \{\alpha \in \mathsf{On} : \mathbb{p} \Vdash_{\mathbb{P}} \text{“} \check{\alpha} \mathrel{\varepsilon} \underset{\sim}{s} \text{”} \text{ for a } \mathbb{p} \in \mathbb{G}\} \text{ for a } \mathbb{P}\text{-name } \underset{\sim}{s} \text{ such that } \Vdash_{\mathbb{P}} \text{“} \underset{\sim}{s} \mathrel{\varepsilon} \underset{\sim}{S} \text{ and } \sup(\underset{\sim}{s}) \equiv \sup(b) \text{”}\}.$$

Note that if $\mathbb{G}$ is a $(\mathsf{V}, \mathbb{P})$-generic filter, then $\underset{\sim}{S}(\mathbb{G}) = \underset{\sim}{S}[\mathbb{G}]$.

For uncountable cardinals μ and $\kappa > \aleph_1$, let $\mathsf{MA}^{++\mu}(\mathcal{P}, < \kappa)$ be the strengthening of $\mathsf{MA}^{+\mu}(\mathcal{P}, < \kappa)$ defined by:

$\mathsf{MA}^{++\mu}(\mathcal{P}, < \kappa)$: *For any $\mathbb{P} \in \mathcal{P}$, any family $\mathcal{D}$ of dense subsets of $\mathbb{P}$ with $|\mathcal{D}| < \kappa$ and any family $\mathcal{S}$ of $\mathbb{P}$-names such that $|\mathcal{S}| \leq \mu$ and $\Vdash_{\mathbb{P}}$ “$\underset{\sim}{S}$ is a stationary subset of $\mathcal{P}_{\eta_{\underset{\sim}{S}}}(\theta_{\underset{\sim}{S}})$” for some $\omega < \eta_{\underset{\sim}{S}} \leq$*

$\theta_{\underset{\sim}{S}} < 2^{\aleph_0}$ *with* $\eta_{\underset{\sim}{S}}$ *regular, for all* $\underset{\sim}{S} \in \mathcal{S}$, *there is a* $\mathcal{D}$*-generic filter* $\mathbb{G}$ *over* $\mathbb{P}$ *such that* $\underset{\sim}{S}(\mathbb{G})$ *is stationary in* $\mathcal{P}_{\eta_{\underset{\sim}{S}}}(\theta_{\underset{\sim}{S}})$ *for all* $\underset{\sim}{S} \in \mathcal{S}$.

In case of $\mu = \omega_1$, the principle $\mathsf{MA}^{++\omega_1}(\mathcal{P}, {<}\kappa)$ is equivalent to the usual $\mathsf{MA}^{+\omega_1}(\mathcal{P}, {<}\kappa)$.

Proposition 6.2. *Suppose that* κ *is a supercompact cardinal and* f *a Laver function on* κ. *Let* S *and* T *be defined by* (5.3) *and* (5.4).

(1) *Suppose that the preparatory finite support ccc iteration* $\langle \mathbb{O}_\alpha, \underset{\sim}{\mathbb{R}}_\beta : \alpha \leq \kappa, \beta < \kappa \rangle$ *is defined by:*

$$\underset{\sim}{\mathbb{R}}_\beta = \begin{cases} (P)^{\checkmark}_{\mathbb{O}_\beta}, \text{ if } \beta = \alpha + 1 \text{ for an } \alpha \in S \text{ and} \\ \qquad f(\alpha) = \langle \mu, \theta, P \rangle \text{ for cardinals } \mu,\, \theta \text{ and a poset } P \\ \qquad \text{such that } \Vdash_{\mathbb{O}_\beta} \text{“}(P)^{\checkmark}_{\mathbb{O}_\beta} \text{ is a ccc poset”}; \\ \{\mathbb{1}_{\underset{\sim}{\mathbb{R}}_\beta}\}, \text{ otherwise.} \end{cases} \tag{6.5}$$

Then, for the Easton-type mixed support iteration $\langle \mathbb{P}_\alpha, \underset{\sim}{\mathbb{Q}}_\beta : \alpha \leq \kappa, \beta < \kappa \rangle$ *over* $\langle \mathbb{O}_\alpha, \underset{\sim}{\mathbb{R}}_\beta : \alpha \leq \kappa, \beta < \kappa \rangle$ *and* $\mathbb{P} = \mathbb{P}_\kappa * (\mathrm{Col}(\kappa, \mu))^{\bullet}_{\mathbb{P}_\kappa}$ *for a regular* $\mu > \kappa$,

$$\Vdash_{\mathbb{P}} \text{“}\mathsf{MA}^{++\eta}(\mathcal{P}, {<}2^{\aleph_0}) \text{ for all cardinal } \eta < 2^{\aleph_0} \text{ holds for } \mathcal{P} = \{P : P \text{ is a ccc poset and } P \in \mathsf{V}\}\text{”}, \tag{6.6}$$

where “P is ccc poset” is meant “ccc poset in the $\mathbb{P}$*-generic extension” while* V *denotes here the ground model before extending generically by* $\mathbb{P}$.

(2) *Suppose that* $\langle \mathbb{O}_\alpha, \underset{\sim}{\mathbb{R}}_\beta : \alpha \leq \kappa, \beta < \kappa \rangle$ *is a preparatory finite support iteration such that each* $\underset{\sim}{\mathbb{Q}}_\alpha$ *for* $\alpha \in T$ *is a* $\mathbb{O}_\alpha$*-name of the Cohen poset* $\mathrm{Fn}(\omega_2)$. *Then, for the Easton-type mixed support iteration* $\langle \mathbb{P}_\alpha, \underset{\sim}{\mathbb{Q}}_\beta : \alpha \leq \kappa, \beta < \kappa \rangle$ *over* $\langle \mathbb{O}_\alpha, \underset{\sim}{\mathbb{R}}_\beta : \alpha \leq \kappa, \beta < \kappa \rangle$ *and* $\mathbb{P} = \mathbb{P}_\kappa * (\mathrm{Col}(\kappa, \mu))^{\bullet}_{\mathbb{P}_\kappa}$ *for a regular* $\mu > \kappa$,

$$\Vdash_{\mathbb{P}} \text{“}\mathsf{MA}^{++\eta}(\mathcal{P}, {<}2^{\aleph_0}) \text{ for all cardinal } \eta < 2^{\aleph_0} \text{ holds for } \mathcal{P} = \{P : P \text{ is forcing equivalent to } \mathrm{Fn}(\mu, 2) \text{ for some } \mu\}\text{”}, \tag{6.7}$$

Proof. (1): Suppose that $\mathbb{G}_\kappa$ is a $(\mathsf{V}, \mathbb{P}_\kappa)$-generic filter and g a $(\mathsf{V}[\mathbb{G}_\kappa], \mathrm{Col}(\kappa, \mu)^{\mathsf{V}[\mathbb{G}_\kappa]})$-generic filter. Let η, $\nu < \kappa$, $\mathbb{R} \in V$, and $\mathcal{D}$, $\mathcal{S} \in V[\mathbb{G}_\kappa * \mathrm{g}]$ be such that

(6.8) $\quad \mathsf{V}[\mathbb{G}_\kappa][\mathbb{g}] \models$ "$\mathbb{R}$ is a ccc poset,
$\mathcal{D}$ is a family of dense subsets of $\mathbb{R}$ with $|\mathcal{D}| = \nu$,
and $\mathcal{S}$ is a family of $\mathbb{R}$-names with $|\mathcal{S}| = \eta$ such
that each element $\underset{\sim}{S}$ of $\mathcal{S}$ is a $\mathbb{R}$-name of a station-
nary subset of $\mathcal{P}_{\eta_{\underset{\sim}{S}}}(\theta_{\underset{\sim}{S}})$ for some $\omega < \eta_{\underset{\sim}{S}} \leq \theta_{\underset{\sim}{S}} < 2^{\aleph_0}$
with $\eta_{\underset{\sim}{S}}$ regular".

Let $|\mathbb{R}| = \lambda$. Without loss of generality, we may assume that the underlying set of $\mathbb{R}$ is λ. Thus $\mathbb{R} = \langle \lambda \leq_{\mathbb{R}} \rangle$. Let θ be sufficiently large and let $j : V \overset{\preceq}{\to} M$ be such that $crit(j) = \kappa$, $j(\kappa) > \theta$,

(6.9) $\quad [M]^\theta \subseteq M$, and

(6.10) $\quad j(f)(\kappa) = \langle \mu, \theta, \mathbb{R} \rangle$.

Let $S^* = j(S)$, $\nu^* = j(\nu)$ and, let $\vec{\mathbb{P}}^* = j(\langle \mathbb{P}_\alpha, \underset{\sim}{\mathbb{Q}}_\beta : \alpha \leq \kappa, \beta < \kappa \rangle)$. As before, we write

(6.11) $\quad \vec{\mathbb{P}}^* = \langle \mathbb{P}^*_\alpha, \underset{\sim}{\mathbb{Q}}^*_\beta : \alpha \leq j(\kappa), \beta < j(\kappa) \rangle$.

We have $\mathbb{P}_\alpha = \mathbb{P}^*_\alpha$ for $\alpha \leq \kappa$.

Let $\mathbb{G}_\kappa$ be a $(\mathsf{V}, \mathbb{P}_\kappa)$-generic filter. Then $\underset{\sim}{\mathbb{Q}}^*_\kappa[\mathbb{G}_\kappa] = \mathrm{Col}(\kappa, \mu)^{\mathsf{V}[\mathbb{G}_\kappa]}$ and $\nu^*(\kappa) \geq \theta$ by (6.10). Let $\mathbb{g}$ be a $(\mathsf{V}[\mathbb{G}_\kappa], \mathrm{Col}(\kappa, \mu)^{\mathsf{V}[\mathbb{G}_\kappa]})$-generic filter. By (6.10) and (6.8), $\underset{\sim}{\mathbb{Q}}_{\kappa+1}[\mathbb{G}_\kappa * \mathbb{g}] \sim \mathbb{R}$.

Thus, in M, $\mathbb{P}^*_{j(\kappa)}$ is factored as

(6.12) $\quad \mathbb{P}^*_{j(\kappa)} \;\sim\; \mathbb{P}_\kappa * \mathrm{Col}(\kappa, \mu)^\bullet_{\mathbb{P}_\kappa} * \underset{\sim}{\mathbb{R}} * \underset{\sim}{\mathbb{R}}_1$

where $\underset{\sim}{\mathbb{R}}$ corresponds to the ground model poset $\mathbb{R}$. We have

(6.13) $\quad \Vdash_{\mathbb{P}_\kappa * \mathrm{Col}(\kappa,\mu)^\bullet_{\mathbb{P}_\kappa} * \underset{\sim}{\mathbb{R}}}$ "$\underset{\sim}{\mathbb{R}}_1$ is a regular sub-poset of the completion of
a poset of the form 'ccc poset $\times\, \nu^*(\kappa)$-closed
poset' "

by (5.10), (5.12), and (5.14).

Now, let $\mathbb{h}$ be $(M[\mathbb{G} * \mathbb{g}], \mathbb{R})$-generic filter and $\mathbb{H}$ be $(M[\mathbb{G} * \mathbb{g} * \mathbb{h}], \underset{\sim}{\mathbb{R}}_1[\mathbb{G} * \mathbb{g} * \mathbb{h}])$-generic filter. $j : \mathsf{V} \overset{\preceq}{\to} M$ is then lifted to

(6.14) $\quad j^* : \mathsf{V}[\mathbb{G}_\kappa] \overset{\preceq}{\to} M[\mathbb{G}_\kappa * \mathbb{g} * \mathbb{h} * \mathbb{H}]$; $\underset{\sim}{a}[\mathbb{G}_\kappa] \mapsto j(\underset{\sim}{a})[\mathbb{G}_\kappa * \mathbb{g} * \mathbb{h} * \mathbb{H}]$.

We have

(6.15) $M \models$ "$\mathbb{C}^* = j^*(\mathrm{Col}(\kappa,\mu)^{\mathsf{V}[\mathbb{G}_\kappa]})$ is $<j(\kappa)$-directed closed"

by elementarity and $(\mu^{<\kappa})^+ < \theta < j(\kappa)$. Thus, we can find an $(M[\mathbb{G}_\kappa * \mathrm{g} * \mathbb{h} * \mathbb{H}], \mathbb{C}^*)$-generic filter g^* such that $j^{*\,\prime\prime}\mathrm{g} \subseteq \mathrm{g}^*$. j^* is then further lifted to

(6.16) $j^{**} : (\mathsf{V}[\mathbb{G}_\kappa])\,[\mathrm{g}] \overset{\preceq}{\to} (M[\mathbb{G}_\kappa * \mathrm{g} * \mathbb{h} * \mathbb{H}])\,[\mathrm{g}^*]$; $\underset{\sim}{a}[\mathrm{g}] \mapsto j^*(\underset{\sim}{a})[\mathrm{g}^*]$

for $\mathrm{Col}(\kappa,\mu)^{\mathsf{V}[\mathbb{G}_\kappa]}$-names $\underset{\sim}{a}$ in $\mathsf{V}[\mathbb{G}_\kappa]$.

In $M[\mathbb{G}_\kappa * \mathrm{g} * \mathbb{h}]$, $\mathbb{h}$ is a filter on $\mathbb{R}$ with intersects with each element of $\mathcal{D}$ and each element $\underset{\sim}{S}$ of $\mathcal{S}$ is interpreted as a stationary subset of $\mathcal{P}_{\eta_{\underset{\sim}{S}}}(\theta_{\underset{\sim}{S}})$ by the genericity of $\mathbb{h}$ and since $\mathcal{D}$, $\mathcal{S} \in M$ by the closedness (6.9) of M. These interpretations of elements of $\mathcal{S}$ remain stationary in $M[\mathbb{G}_\kappa * \mathrm{g} * \mathbb{h} * \mathbb{H} * \mathrm{g}^*]$ by (6.13). and (6.15) (see Lemma 3.1).

Since $\mathcal{D}$ and $\mathcal{S}$ have cardinality $< \kappa$, we have

(6.17) $j^{**}(\mathcal{D}) = \{j^{**}(D) : D \in \mathcal{D}\}$, and
$j^{**}(\mathcal{S}) = \{j^{**}(\underset{\sim}{S}) : \underset{\sim}{S} \in \mathcal{S}\}$.

It follows that $\mathbb{h}$ generates a filter on $j^{**}(\mathbb{R})$ which intersects each element of $j^{**}(\mathcal{D})$ and interprets each element $\underset{\sim}{S}$ of $j^{**}(\mathcal{S})$ as a superset of the corresponding interpretation of the element of $\mathcal{S}$ by $\mathbb{h}$ is a stationary subset of $\mathcal{P}_{\eta_{\underset{\sim}{S}}}()$ in $M[\mathbb{G}_\kappa * \mathrm{g} * \mathbb{h} * \mathbb{H} * \mathrm{g}^*]$.

Thus, we have

(6.18) $M[\mathbb{G}_\kappa * \mathrm{g} * \mathbb{h} * \mathbb{H} * \mathrm{g}^*] \models$
"there is a $j^{**}(\mathcal{D})$-generic filter on $j^{**}(\mathbb{R})$
which interprets each element $\underset{\sim}{S}$ of $j^{**}(\underset{\sim}{S})$
as a stationary subset of $\mathcal{P}_{\eta_{\underset{\sim}{S}}}(\theta_{\underset{\sim}{S}})$".

By the elementarity of j^{**}, it follows that

(6.19) $\mathsf{V}[\mathbb{G}_\kappa * \mathrm{g}] \models$ "there is a $\mathcal{D}$-generic filter on $\mathbb{R}$
which interprets each element $\underset{\sim}{S}$ of $\underset{\sim}{S}$
as a stationary subset of $\mathcal{P}_{\eta_{\underset{\sim}{S}}}(\theta_{\underset{\sim}{S}})$".

(2): can be proved similarly to (1). □ (Proposition 6.2)

Theorem 6.3. (1) *Suppose that the existence of two supercompact cardinals is consistent. Then the following combination of the principles is also consistent:*

(6.2) $\mathsf{SDLS}^{int}_{+}(\mathcal{L}^{\aleph_0}_{stat}, <2^{\aleph_0})$, $\mathsf{GRP}^{<\kappa}(\leq 2^{\aleph_0})$;

(6.3) $\mathsf{SDLS}^{int}_{+}(\mathcal{L}^{\mathsf{PKL}}_{stat}, <2^{\aleph_0})$;

(6.20) $\mathsf{MA}^{++\eta}(\mathcal{P}, <2^{\aleph_0})$ *for* $\mathcal{P} = \{\mathbb{P} : \mathbb{P} \sim \mathrm{Fn}(\lambda, 2)$ *for some* $\lambda\}$ *for all* $\eta < 2^{\aleph_0}$*; and*

(6.21) $\mathsf{HH}(<2^{\aleph_0})$.

(2) *If there is a superhuge cardinal and a supercompact cardinal above it, then the combination of the principles (6.2) ∼ (6.21) above together with*

(6.22) $\mathcal{P}_{2^{\aleph_0}}(\lambda)$ *carries a* σ*-saturated normal ideal for all* $\lambda \geq 2^{\aleph_0}$

is consistent.

Proof. (1): For two supercompact cardinals $\kappa < \kappa_1$, let $\langle \mathbb{O}_\alpha, \underset{\sim}{\mathbb{R}}_\beta : \alpha \leq \kappa, \beta < \kappa\rangle$ and $\langle \mathbb{P}_\alpha, \underset{\sim}{\mathbb{Q}}_\beta : \alpha \leq \kappa, \beta < \kappa\rangle$ be as in Proposition 6.2, (2).

Then, the generic extension of V by $\mathbb{P} = \mathbb{P}_\kappa * (\mathrm{Col}(\kappa, \kappa_1))^{\bullet}_{\mathbb{P}_\kappa}$ is as desired: $\Vdash_{\mathbb{P}}$ "(6.1), (6.2), (6.3)" by Theorem 6.1. $\Vdash_{\mathbb{P}}$ "(6.20)" by Proposition 6.2,(2) and $\Vdash_{\mathbb{P}}$ "(6.21)" by Theorem 5.16, (2) and Proposition 4.7.

(2): Let $\mathbb{P} = \mathbb{P}_\kappa * (\mathrm{Col}(\kappa, \kappa_1))^{\bullet}_{\mathbb{P}_\kappa}$ be as in (1) but for a superhuge κ and a supercompact κ_1 above κ. Then $\Vdash_{\mathbb{P}}$ "(6.2) ∼ (6.21)" as in (1) and $\Vdash_{\mathbb{P}}$ "(6.22)" by Theorem 5.16 ,(2′) and Proposition 4.9. ❑ (Theorem 6.3)

Theorem 6.4. (1) *Suppose that the existence of two supercompact cardinals is consistent. Then the following combination of principles is also consistent:*

(6.2) $\mathsf{SDLS}^{int}_{+}(\mathcal{L}^{\aleph_0}_{stat}, <2^{\aleph_0})$, $\mathsf{GRP}^{<\kappa}(\leq 2^{\aleph_0})$,

(6.3) $\mathsf{SDLS}^{int}_{+}(\mathcal{L}^{\mathsf{PKL}}_{stat}, <2^{\aleph_0})$;

(6.23) *There is an inner model* M *of* V *such that* $(2^{\aleph_0})^{\mathsf{V}} = (2^{\aleph_0})^M$ *and* V *is reached from* M *by the forcing with a regular sub-poset of the completion of the product of ccc and* $<2^{\aleph_0}$*-closed posets, and* $\mathsf{MA}^{++\eta}(\mathcal{P}, <2^{\aleph_0})$ *for* $\mathcal{P} = \{\mathbb{P} : \mathbb{P}$ *is a ccc poset* $\mathbb{P} \in M\}$ *for all* $\eta < 2^{\aleph_0}$*, and*

(6.24) $\neg\,\mathsf{HH}(<2^{\aleph_0})$.

(2) *If there is a superhuge cardinal and a supercompact cardinal above it, then the combination of the principles* (6.2), (6.3), (6.23) *and* (6.24) *above together with*

(6.22) $\mathcal{P}_{2^{\aleph_0}}(\lambda)$ *carries a* σ*-saturated normal ideal for all* $\lambda \geq 2^{\aleph_0}$

is consistent.

Proof. (1): Let $\langle \mathbb{O}_\alpha, \underset{\sim}{\mathbb{R}}_\beta : \alpha \leq \kappa, \beta < \kappa \rangle$ be the following modification of the preparatory ccc finite support iteration (6.5) in Proposition 6.2, (1). Then:

$$(6.25) \quad \underset{\sim}{\mathbb{R}}_\beta = \begin{cases} (P)^{\checkmark}_{\mathbb{O}_\beta}, & \text{if } \beta = \alpha+1 \text{ for an } \alpha \in S \text{ and} \\ & f(\alpha) = \langle \mu, \theta, P \rangle \text{ for cardinals } \mu, \theta \text{ and a poset } P \\ & \text{such that } \Vdash_{\mathbb{O}_\beta} \text{“}(P)^{\checkmark}_{\mathbb{O}_\beta} \text{ is a ccc poset”;} \\ \text{Hechler real forcing over } \mathbb{O}_\beta, & \text{if } \beta \in T \text{ but} \\ & \beta \text{ is not a successor of an element of } S; \\ \{\mathbb{1}_{\underset{\sim}{\mathbb{R}}_\beta}\}, & \text{otherwise.} \end{cases}$$

Let $\langle \mathbb{P}_\alpha, \underset{\sim}{\mathbb{Q}}_\beta : \alpha \leq \kappa, \beta < \kappa \rangle$ be the Easton-type mixed support iteration over $\langle \mathbb{O}_\alpha, \underset{\sim}{\mathbb{R}}_\beta : \alpha \leq \kappa, \beta < \kappa \rangle$ and $\mathbb{P} = \mathbb{P}_\kappa * (\mathrm{Col}(\kappa, \kappa_1))^\bullet_{\mathbb{P}_\kappa}$.

Then, the generic extension of V by $\mathbb{P} = \mathbb{P}_\kappa * (\mathrm{Col}(\kappa, \kappa_1))^\bullet_{\mathbb{P}_\kappa}$ is as desired: $\Vdash_{\mathbb{P}}$ “(6.1), (6.2), (6.3)” by Theorem 6.1.

$\Vdash_{\mathbb{P}}$ “(6.23)” follows from (the proof of) Proposition 6.2, (1). Note that the proof of Proposition 6.2, (1) does not rely on the value of $\mathbb{R}_\beta$ for $\beta \in T$ which is not a successor of the element of S.

Now the Hechler part of the preparatory iteration introduces an $\leq^*$-increasing sequence $\vec{h} = \langle f_\alpha : \alpha < 2^{\aleph_0} \rangle$ of functions of length $2^{\aleph_0}$ (in the generic extension by $\mathbb{P}_\kappa$) $\mathcal{F} = \{f_\alpha : \alpha < 2^{\aleph_0}\}$ is still unbounded in the generic extension by $\mathbb{P}_\kappa$ by the genericity of f_α's: f_α's may no more Hechler reals above corresponding intermediate models in $\mathsf{V}^{\mathbb{P}_\kappa}$ but each of them adds a Cohen real as its coordinatewise summand (see [Truss[21]]). $\mathcal{F}$ remains unbounded in $\mathbb{P}$-generic extension $\mathsf{V}[\mathbb{G}_\kappa * \mathbb{H}]$ since no new reals are added by $\mathrm{Col}(\kappa, \kappa_1)$. Thus, in $\mathsf{V}[\mathbb{G}]$, the first countable topological space $X_{\mathcal{F}}$ constructed in Section 2 is non-metrizable but all subspaces of $X_{\mathcal{F}}$ of size $< 2^{\aleph_0}$ are metrizable. Thus $\Vdash_{\mathbb{P}}$ “(6.24)”.

(2): Let $\mathbb{P} = \mathbb{P}_\kappa * (\mathrm{Col}(\kappa, \kappa_1))^\bullet_{\mathbb{P}_\kappa}$ be as in (1) with superhuge κ and supercompact κ_1 above κ. Then $\Vdash_{\mathbb{P}}$ “(6.2), (6.3), (6.23), (6.24) ” as in (1) and $\Vdash_{\mathbb{P}}$ “(6.22)” by Theorem 5.16, (2′) and Proposition 4.9. □ (Theorem 6.4)

We end up with mentioning some remaining open problems. As noted in Section 2, Hamburger's Problem i.e. the consistency of $\mathsf{HH}(< \aleph_2)$ is still widely open. Galvin's Conjecture is also a persistently open problem which can be discussed in our context (see e.g. [Todorcevic[20]]).

Both of the following two problems, which might be more at hand, are related to the last theorem in this section:

Problem 6.5. *Can we have the full* $\mathsf{MA}^{++\mu}(ccc, <2^{\aleph_0})$ *together with all other strong reflection properties in some modification of the model of Theorem 6.4?*

Problem 6.6. *What is the* $\mathfrak{Refl}_{\mathsf{HH}}$ *in the model of Theorem 6.4? Can we make it* $(2^{\aleph_0})^+$ *or* ∞ *by some modification of the construction in the proof?*

References

1. Joan Bagaria and Menachem Magidor, On ω_1-strongly compact cardinals, Vol.79, (1), (2014), 266–278.
2. R.H. Bing, Metrization of topological spaces, Canadian Journal of Mathematics Vol.3 (1951), 175–186.
3. Sean Cox, The diagonal reflection principle, Proceedings of the American Mathematical Society, Vol.140, No.8 (2012), 2893-2902.
4. James Cummings, Iterated Forcing and Elementary Embeddings, in: (Matthew Foreman and Akihiro Kanamori, eds.) Handbook of set Theory, Vol.2, (2009), 775–884.
5. Eric K. van Douwen, The integers and topology, in: K. Kunen and J. Vaughan (eds.), Handbook of Set-Theoretic Topology, Elsevier, (1984).
6. Alan Dow, An introduction to applications of elementary submodels to topology, Topology Proceedings 13, No.1 (1988), 17–72.
7. A. Dow, F.D.Tall, and W.A.R. Weiss, New proofs of the consistency of the normal Moore space conjecture I, II, Topology and its Applications, 37 (1990), 33–51, 115–129.
8. Sakaé Fuchino, Istvan Juhász, Lajos Soukup, Zoltán Szentmiklóssy, and Toshimichi Usuba, Fodor-type Reflection Principle and reflection of metrizability and meta-Lindelöfness, Topology and its Applications, Vol.157, 8 (2010), 1415–1429.
9. Sakaé Fuchino, André Ottenbereit Maschio Rodrigues, and Hiroshi Sakai, Strong downward Löwenheim-Skolem theorems for stationary logics, I, to appear in Archive for Mathematical Logic.
   Extended version of the paper: `https://fuchino.ddo.jp/papers/SDLS-x.pdf`
10. ________________: Strong downward Löwenheim-Skolem theorems for stationary logics, II — reflection down to the continuum, preprint.
   Extended version of the paper: `https://fuchino.ddo.jp/papers/SDLS-II-x.pdf`
11. Sakaé Fuchino and André Ottenbreit Maschio Rodrigues, Reflection principles, generic large cardinals, and the Continuum Problem, to appear in the Proceedings of the Symposium on Advances in Mathematical Logic 2018.
   Extended version of the paper:
   `https://fuchino.ddo.jp/papers/refl_principles_gen_large_cardinals_continuum_problem-x.pdf`
12. A. Hajnal and I. Juhász, On spaces in which every small subspace is metrizable, Bulletin de lAcadémie Polonaise des Sciences, Série des sciences math, astr. et phys, Vol.24, No9, (1976), 727–731.

13. Thomas Jech, Theory, The Third Millennium Edition, Springer (2001/2006).
14. Akihiro Kanamori, *The Higher Infinite*, Second Edition, Springer Monographs in Mathematics, Springer-Verlag, (2003/2009).
15. John Krueger, A general Mitchell style iteration, Mathematical Logic Quarterly, 54(6), (2008), 641–651.
16. ___________, Some applications of mixed support iterations, Annals of Pure and Applied Logic, 158(1-2), (2009), 40–57.
17. K. Kunen, *Set Theory, An Introduction to Independence Proofs*, North-Holland (1980).
18. André Ottenbreit Maschio Rodrigues, Some reflection principles in generic extensions by mixed support iteration, PhD thesis, Kobe University (2020).
19. Hiroshi Sakai, Posets of cardinality κ which destroy stationary subsets of $\mathcal{P}_\kappa(\lambda)$, in preparation.
20. Stevo Todorcevic, Combinatorial dichotomies in set theory, The Bulletin of Symbolic Logic, Vol.17, No.1, (2011), 1–72.
21. John Truss, Sets having calibre $\aleph_1$, in: R.Gandy, M.Hyland (Eds.), Logic Colloquium 76 , North-Holland PUblishing Company (1977), 595–612.

On interior of definable subsets of ordered o-ω-stable fields

V. V. Verbovskiy

Satbayev University, 22a Satbayev St.,
Almaty, 050013, The Republic of Kazakhstan
E-mail: viktor.verbovskiy@gmail.com

An ordered structure $\mathcal{M}$ is called o-λ-stable if for any subset A with $|A| \le \lambda$ and any cut in $\mathcal{M}$ there are at most λ 1-types over A which are consistent with the cut. In this article, we prove that any infinite definable subset of an ordered o-ω-stable field contains a non-empty interior, that is, it is a union of an open set and a finite set.

Keywords: Model theory, ordered fields, o-minimal, o-stable, NIP.

1. Introduction

B. Baizhanov and V. Verbovskiy[1] introduced a notion of an o-stable theory, that allows to apply the machinery of the stability theory to investigating ordered structures. In[8], the author considered some properties of ordered groups and fields that have an o-stable elementary theory, in particular, he proved that such a group is Abelian and such a field without non-trivial definable convex subgroups of the additive group is real closed.

By definition of o-minimality any definable infinite subset of an o-minimal structure has a non-empty interior. The same holds for weakly o-minimal structures. In[6], P. Simon proved that any definable infinite subset of a divisible ordered dp-minimal group has a non-empty interior. For an ordered group with an o-stable theory this fact in general is not true. Let us consider the ordered group of reals $(\mathbb{R}, <, +, Q)$, where Q distinguishes the set of rational numbers. From[8] it follows that the elementary theory of this structure is o-stable, nevertheless Q defines the infinite set which has an empty interior. In[9], the author proved that any ordered group, whose elementary theory has o-Morley rank 1 and which has boundedly many definable convex subgroups, is weakly o-minimal; this implies that any infinite definable subset of a such group has a non-empty interior.

This research aims to prove the hypothesis that an ordered field with o-ω-stable theory is real closed. As it has been shown in[3] there exists a dp-minimal ordered field which is not real closed, for instance, $\mathbb{R}((t^{\Gamma}))$, where Γ is a non-singular ordered abelian group. The author proved in[7] that any dp-minimal theory with a linear order is o-stable. From results by P. Simon[6] it follows that any dp-minimal theory with a linear order is o-superstable. So, in difference with the stability theory where a superstable field is algebraically closed, there exists an o-superstable ordered field which is not real closed. In order to prove that an o-ω-stable ordered field is real closed we can follow the approach by D. Macpherson, D. Marker, and C. Steinhorn[4]. This paper consists of the first step in this direction—of proving that any definable infinite subset of an ordered field whose elementary theory is o-ω-stable contains a non-empty interior, or by the other words, contains a non-empty interval.

Let $\mathcal{M} = (M, <, \dots)$ be a totally ordered structure, $a \in M$, and let A, $B \subseteq M$. As usually we write $a < A$ if $a < b$ for any $b \in A$, and $A < B$ if $a < b$ for any $a \in A$ and $b \in B$. A partition $\langle C, D\rangle$ of M is called a *cut* if $C < D$. Given a cut $\langle C, D\rangle$ we construct a partial type $\{c < x < d : c \in C,\ d \in D\}$, which we also call a cut and use the same notation $\langle C, D\rangle$. If the set C is definable, then the cut is called also *definable.*

We denote the set of all definable in $\mathcal{M}$ cuts $\langle C, D\rangle$, such that D has no smallest element, by $\overline{M}$. The set M can be regarded as a subset of $\overline{M}$ by identifying an element $a \in M$ with the cut $\langle(-\infty, a], (a, +\infty)\rangle$. After such an identification, $\overline{M}$ is naturally equipped with a linear ordering extending $(M, <)$: $\langle C_1, D_1\rangle \le \langle C_2, D_2\rangle$ iff $C_1 \subseteq C_2$. Clearly, $(M, <)$ is a dense substructure of $(\overline{M}, <)$.

If $C = M$ we denote the cut $\langle C, \emptyset\rangle$ as $+\infty$ and $\sup A$ as $\langle C, D\rangle$, where $C = \{c \in M : c < a \text{ for some } a \in A\}$. So, we write $b < \sup A$ if there exists $a \in A$ such that $b < a$. A cut $\langle C, D\rangle$ in an ordered Abelian group is called *non-valuational*[4,10] if for any positive element ε there exist $c \in C$ and $d \in D$ such that $d - c < \varepsilon$. A cut, which is not non-valuational, is called *valuational.* Observe that for a valuational cut $\langle C, D\rangle$ there is a convex non-trivial subgroup H such that $\sup C = \sup(a + H)$ for some a, and this cut is definable if and only if H is definable. An ordered group G is said to be of *non-valuational type*, if any definable cut is non-valuational. Note that G is of non-valuational type if and only if there is no definable non-trivial convex subgroup in G.[4]

A subset A of M is called *convex* if $[a, b] \subseteq A$ for any a and $b \in A$. The *length* of a convex subset A of an ordered Abelian group is defined as $\sup\{a - b : a, b \in A\}$. A *convex component* of a set A is a maximal convex subset of A. The *convex hull* A^c of A is defined as $A^c = \{b \in M : \exists a_1, a_2 \in A(a_1 \leq b \leq a_2)\}$. An ordered structure is called *weakly o-minimal* if any definable subset consists of finitely many convex components[2,4].

Let A be a subset of a totally ordered structure M, and let P be some property. We say that P holds *eventually in* A if there is an element $a \in M$ such that $a < \sup A$ and P holds on $(a, \infty) \cap A$. If $A = M$, we omit it. If sets B and C are eventually equal in A we denote this by $B \overset{\infty}{=}_A C$. Let $B \subseteq A \subseteq M$. The set B is said to be *dense in* A if for any $a_1 < a_2$ from A there is $b \in B$ with $a_1 < b < a_2$. If $A = M$ we omit it. A *dense component* of B in A is a maximal subset B_0 of B which is dense in $A \cap (\inf B_0, \sup B_0)$.

Let T be an $\mathcal{L}$-theory and $\phi(\bar{x}, \bar{y})$ a formula. A formula ϕ has the *independence property* (relative to T) if for each $n < \omega$ there is a model $\mathcal{M} \models T$ and two sequences $(\bar{a}_i : i < n)$ and $(\bar{b}_J : J \subseteq n)$ in M such that $\mathcal{M} \models \phi(\bar{a}_i, \bar{b}_J)$ if and only if $i \in J$. A theory T has the independence property if some formula has the independence property relative to T.

Let s be a partial n-type, A a set. Then

$$S^n_s(A) := \{p \in S^n(A) : p \cup s \text{ is consistent}\}.$$

Note, s need not to be a partial type over A.

Definition 1.1.[1]

(1) An ordered structure $\mathcal{M}$ is *o-stable in* λ if for any $A \subseteq M$ with $|A| \leq \lambda$ and for any cut $\langle C, D \rangle$ in $\mathcal{M}$ there are at most λ one-types over A which are consistent with the cut $\langle C, D \rangle$, i.e. $|S^1_{\langle C,D \rangle}(A)| \leq \lambda$.
(2) A theory T is *o-stable in* λ if every model of T is. Sometimes we write T is o-λ-stable.
(3) T is *o-stable* if there exists a λ in which T is o-stable.

Note that an o-ω-stable theory is o-stable in λ for any infinite λ.[8]

Definition 1.2.[8] A totally ordered structure $\mathcal{M} = (M, <, \dots)$ has *the strict order property inside a cut*, if there is a formula $\phi(x, \bar{y})$ such that for any n there are $\bar{a}_1, \dots, \bar{a}_n$ and a cut $\langle C_n, D_n \rangle$ in $\mathcal{M}$ such that in some $|M|^+$-saturated elementary extension $\mathcal{N}$ of $\mathcal{M}$ the following holds:

$$\phi(\mathcal{N}, \bar{a}_1) \cap \langle C_n, D_n \rangle(\mathcal{N}) \subset \phi(\mathcal{N}, \bar{a}_2) \cap \langle C_n, D_n \rangle(\mathcal{N}) \subset \\ \cdots \subset \phi(\mathcal{N}, \bar{a}_n) \cap \langle C_n, D_n \rangle(\mathcal{N}).$$

A theory is said to have the *strict order property inside a cut* if it holds in some of its models.

The following fact is an analog of the fact that a stable theory does not have the strict order property.

Fact 1.1.[8] Any o-stable theory does not have the strict order property inside a cut.

Fact 1.2.[1] An o-stable theory does not have the independence property.

Facts 1.1 and 1.2 implies the following criterion of o-stability.

Fact 1.3.[8] Let a language L contain binary symbol '$<$' and a theory T of L which includes an axiom saying that '$<$' is a total order. The theory T is o-stable if and only if it does not have the independence property and the strict order property inside a cut.

Definition 1.3.[8]

(1) We say that Morley o-rank of a formula $\phi(x)$ inside a cut $\langle C, D\rangle$ in $\mathcal{M}$ is equal to or greater than 1 and write $RM_{\langle C,D\rangle,\mathcal{M}}(\phi)\geq 1$ for this, if $\{\phi(x)\} \cup \langle C, D\rangle$ is consistent.
(2) $RM_{\langle C,D\rangle,\mathcal{M}}(\phi) \geq \alpha+1$ if there are infinitely many pairwise inconsistent formulae $\psi_i(x)$ such that $RM_{\langle C,D\rangle,\mathcal{M}}(\phi(x) \wedge \psi_i(x)) \geq \alpha$.
(3) If α is a limit ordinal, then $RM_{\langle C,D\rangle,\mathcal{M}}(\phi) \geq \alpha$ if $RM_{\langle C,D\rangle,\mathcal{M}}(\phi) \geq \beta$ for all $\beta < \alpha$.
(4) $RM_{\langle C,D\rangle,\mathcal{M}}(\phi) = \alpha$ if $RM_{\langle C,D\rangle,\mathcal{M}}(\phi) \geq \alpha$ and $RM_{\langle C,D\rangle,\mathcal{M}}(\phi) \not\geq \alpha+1$.
(5) $RM_{\langle C,D\rangle,\mathcal{M}}(\phi) = \infty$ if $RM_{\langle C,D\rangle,\mathcal{M}}(\phi) \geq \alpha$ for any α.
(6) We define Morley o-rank of a formula $\phi(x)$ as follows:
$\text{o-}RM(\phi) = \sup\{RM_{\langle C,D\rangle,\mathcal{M}}(\phi) : \mathcal{M} \models T \text{ and } \langle C, D\rangle \text{ is a cut in } \mathcal{M}\}$.

In a similar way one can define Morley o-degree of a formula inside a cut. As usual one can define Morley o-rank of a type. Note that in an o-ω-stable theory Morley o-rank of any formula is less than ∞.

Throughout this section G is a sufficiently saturated ordered group whose elementary theory is o-ω-stable. Sometimes we write simply that G is o-ω-stable meaning that the elementary theory of G is o-ω-stable. Since any ordered o-stable group is Abellian[8], we use the additive notation. So, we consider an ordered group in some language L which expands the language of ordered abelian groups $\{=, <, +, 0\}$. We say that a formula $\varphi \in L(G)$ is *eventually minimal inside a cut* s if φ is consistent with s and for any formula ψ exactly one of $\varphi \wedge \psi$ and $\varphi \wedge \neg\psi$ is consistent with s. We define the eventual stabilizer of a formula φ as follows:

$$K_\varphi(x) := \exists z \forall y \big[z < y \to (\varphi(y) \leftrightarrow \varphi(y + x))\big]$$

Obviously, $K_\varphi := K_\varphi(G)$ is a subgroup of G.

Note that for any definable convex set A each of $\inf A$ and $\sup A$ defines a convex subgroup H_A^- and H_A^+, respectively, in the following way:

$$H_A^- := \{g \in G : a - |g| \in A \text{ for any } a \in A\},$$
$$H_A^+ := \{g \in G : a + |g| \in A \text{ for any } a \in A\}.$$

Fact 1.4. [8] Let G be a densely ordered o-stable group. Then for any infinite definable set there is an interval where this set is dense. In other words, no infinite nowhere dense subset is definable.

Lemma 1.1. *Let H be a convex non-zero subgroup of an ordered o-ω-stable group G. Let A be a definable unbounded in H subset of H. Then there exists a formula θ, such that $\theta(G) \subseteq A$ and for any element $g \in H$ it holds that*

$$\text{either } \theta(G) \overset{\infty}{=} (\theta(G) + g), \text{ or } \theta(G) \cap (\theta(G) + g) \overset{\infty}{=} \emptyset.$$

Proof. If Morley o-rank of a formula in the cut $\langle \{g < \sup H\}, \{g > \sup H\} \rangle$ is less than infinity, then there exists an unbounded formula ψ of Morley o-rank 1 and Morley o-degree 1. We can take this formula ψ as θ.

If Morley o-rank of a formula is infinity it is easy to construct continuum many types extending the cut $\langle \{g < \sup H\}, \{g > \sup H\} \rangle$ over a countable subset of an appropriate elementary extension of G by the standard machinery of the stability theory, contradicting o-ω-stability of $Th(G)$. □

Fact 1.5. [8] Let φ be such that for any $g \in G$ either $\varphi(G) \overset{\infty}{=} (\varphi(G) + g)$ or $\varphi(G) \cap (\varphi(G) + g) \overset{\infty}{=} \emptyset$. Then K_φ is unbounded.

Fact 1.6. [8] Let G be an o-stable ordered group with boundedly many definable convex subgroups. Assume that G is not weakly o-minimal, that is there is a definable subset A consisting of infinitely many convex components and a non-rational cut $\langle C, D \rangle$ such that both A and the complement of A are consistent with this cut. Then there is an externally definable unbounded proper subgroup K of H_C^+ (where H_C^+ is the stabilizer of the set C). If in addition the cut $\langle C, D \rangle$ is definable then K is definable.

S. Shelah[5] considered the expansion of a model of a NIP theory by the externally definable subsets in order to prove that the elementary theory of the expanded model is NIP. Let T be a theory of a language $\mathcal{L}$, and $\mathcal{M} \prec \mathcal{N}$ two models of T such that $\mathcal{N}$ is $|M|^+$-saturated. For any formula $\phi(\bar{x}, \bar{\alpha})$ with parameters $\bar{\alpha}$ in N we add a new relational symbol $P_{\phi(\bar{x},\bar{\alpha})}(\bar{x})$

interpreted by $P_{\phi(\bar{x},\bar{\alpha})}(M) = \phi(N,\bar{\alpha}) \cap M^k$ in order to form language $\mathcal{L}^*$. We denote the elementary theory of $(M,\mathcal{L}^*)$ by T^*.

Fact 1.7. [8] Let T be an o-stable theory of a language $\mathcal{L}$, and $\mathcal{M} \prec \mathcal{N}$ two models of T such that $\mathcal{N}$ is $|M|^+$-saturated. Then the elementary theory T^* of the expansion $\mathcal{M}^*$ of $\mathcal{M}$ is o-stable.

2. Ordered o-ω-stable fields

We consider a sufficiently saturated expansion of an ordered field $\mathcal{F} = \langle F, <, +, \cdot, 0, 1, \ldots\rangle$ with an o-ω-stable theory. Recall some facts from[8].

Fact 2.1. [8] Any definable subgroup of the additive group of $\mathcal{F}$ is convex.

We say that G has *boundedly many definable convex subgroups* if there is a cardinal λ such that in any group which is elementary equivalent to G the number of convex definable subgroups does not exceed λ. Otherwise we say that G has *unboundedly many definable convex subgroups.*

Fact 2.2. [8] Let $\mathcal{F}$ be an o-stable ordered field with boundedly many convex definable subgroups of the additive group. Then $\mathcal{F}$ is weakly o-minimal and real closed.

Note that in an ordered field there are at most boundedly many definable convex subgroups if and only if all definable convex subgroups are trivial, because multiplying a bounded convex subgroup by an element which is not in this subgroup gives another definable subgroup.

Theorem 2.3. *Any infinite definable subset of an o-ω-stable ordered field contains a non-empty interior.*

Proof. Let $\mathcal{F}$ be an ordered field with an o-ω-stable theory. Let also $\mathcal{F}$ be sufficiently saturated. Assume that there exists an infinite set A definable in $\mathcal{F}$ by a formula $\varphi(x)$, such that the interior of A is empty. By Fact 1.4 there exists an infinite interval (a,b), such that $A \cap (a,b)$ is everywhere dense in (a,b). Let $C = \{(c-a)^{-1} : c \in (a,b) \cap A\}$. Clearly, C is a definable unbounded subset of $\mathcal{F}$. By Lemma 1.1 there exists a formula θ, such that $\theta(\mathcal{F}) \subseteq C$ and for any element $g \in \mathcal{F}$ it holds that either $\theta(\mathcal{F}) \overset{\infty}{=} (\theta(\mathcal{F}) + g)$, or $\theta(\mathcal{F}) \cap (\theta(\mathcal{F}) + g) \overset{\infty}{=} \emptyset$. By Lemma 1.5 the eventual stabilizer K_B of $B = \theta(\mathcal{F})$ is not bounded in $\mathcal{F}$.

If $K_B \neq \mathcal{F}$, then K_B is not convex because K_B is unbounded in $\mathcal{F}$. Since $\mathcal{F}$ has an o-stable theory, there is a contradiction against Fact 2.1. Thus, $K_B = \mathcal{F}$.

Now we use the following fact.

Fact 2.4. [8] Let K_φ be the eventual stabilizer of an arbitrary formula $\varphi(x)$. Assume that K_φ contains a positive element. Then for any $g \in K_\varphi$ there is a non-zero convex in K_φ subgroup H of K_φ such that $g \in H$ and eventually $\varphi(G)$ is equal to $\varphi(G) + H$ (here, H need not be definable).

Let $g \in \mathcal{F}$, and let H be a convex subgroup of the additive group of $\mathcal{F}$, which contains g and satisfies the following: $\theta(\mathcal{F})$ and $\theta(\mathcal{F}) + H$ eventually are equal. By definition there is some $d \in \mathcal{F}$ such that $\theta(\mathcal{F}) \cap (d, \infty) = (\theta(\mathcal{F}) + H) \cap (d, \infty)$. Let $d_1 \in \theta(\mathcal{F}) \cap (d, \infty)$. Then

$$(d_1, d_1 + g) \subseteq (\theta(\mathcal{F}) + H) \cap (d, \infty) = \theta(\mathcal{F}) \cap (d, \infty) \subseteq C.$$

Obviously, $((d_1 + g)^{-1}, d_1^{-1}) \subseteq A$. Thus the interior of A is not empty. □

Lemma 2.1. *Any definable infinite subset of an ordered field with an o-ω-stable theory is a union of an open set and a finite set.*

Proof. Let A be an infinite subset, which is defined by a formula $\varphi(x)$. We write a formula, which names the interior of A.

$$\psi(x) := \exists y \exists z (y < x < z \wedge \forall t (y < t < z \rightarrow \varphi(t))).$$

Note that the set of all realizations of $\psi(x)$ is open.

Now we consider $\theta(x) := \varphi(x) \wedge \neg\psi(x)$. This formula defines all those elements in A which are not in the interior of A. Note that the interior of the set of all realizations of $\theta(x)$ is empty, so this set cannot be infinite. Then A is a union of an open set and a finite set. □

Lemma 2.2. *Let $\mathcal{F}$ be an ordered field whose elementary theory is o-ω-stable. Let A be some definable subset and b some element. Let H denote the least convex subgroup of the additive group of $\mathcal{F}$, which contains b. Then there exists such an element $h \in H$, that either $A \cap (h, \infty) \cap H = \emptyset$, or $A \cap (h, \infty) \cap H = (h, \infty) \cap H$.*

Proof. Assume that a definable set $A \cap H$ is not bounded in H. By Lemma 1.1 there exists a formula θ such that the definable subset $B = \theta(\mathcal{F})$ satisfies the hypotheses of Fact 1.5. Observe that H is definable in the expansion L^* by externally definable subsets, and that T^* is o-stable. We can continue to work in the reduct of T^* to the definable subset H. Consequently, the eventual stabilizer K_B of B in H is not bounded in H. By Fact 2.1 K_B is a convex subgroup that is equal to H. Then h is in K_B. Now we apply Fact 2.4 and obtain that the following identity holds eventually in

H: $B+H \stackrel{\infty}{=}_H B$. Recall that $B+H=H$, then $B\cap(h,\infty)\cap H=(h,\infty)\cap H$. Now we come back to $\mathcal{F}$. Since $B\subseteq A$, so $A\cap(h,\infty)\cap H=(h,\infty)\cap H$ holds. □

Acknowledgments

This research has been/was/is funded by the Science Committee of the Ministry of Science and Higher Education of the Republic of Kazakhstan (Grant No. BR20281002). The author is grateful to the referee for fruitful remarks.

References

1. B. Baizhanov B., V. Verbovskiy, O-stable theories, *Algebra and Logic* **50**, 3 (2011), pp. 303–325.
2. M. Dickmann, Elimination of quantifiers for ordered valution rings, *Proceedings 3rd Easter Model Theory Conference at Gross Koris*, Berlin, 1985.
3. F. Jahnle, P. Simon, and E. Walsberg. Dp-minimal valued fields, *The Journal of Symbolic Logic* **82**, 1 (2017), pp. 151-165.
4. D. Macpherson, D. Marker, and C. Steinhorn, Weakly o-minimal structures and real closed fields, *Trans. Amer. Math. Soc.* **352** (2000), pp. 5435–5483.
5. S. Shelah, Dependent first order theories, continued, *Israel J. Math.* **173** (2009).
6. P. Simon, On dp-minimal ordered strucutres, *The Journal of Symbolic Logic* **76**, 2 (2011), pp. 448–460.
7. V.V. Verbovskiy. Dp-minimalnye i uporyadochenno stabilnye uporyadochennye structury, Matematicheskiy zhurnal, 10 (2010), no 2 (36), 35–38.
8. V. Verbovskiy, O-stable ordered groups, *Siberian Advances in Mathemtics*, **22**, 1 (2012), pp. 60–74.
9. V. Verbovskiy, On ordered groups of Morley o-rank 1, *Siberian Electronic Mathematical Reports*, **15** (2018), pp. 314–320.
10. R. Wencel, Weakly o-minimal non-valuational structures, *Annals of Pure and Applied Logic*, **154**, 3 (2008), pp. 139–162.

On Diophantine equations over $\mathbb{Z}[i]$ with 52 unknowns

Yuri Matiyasevich

St. Petersburg Department of Steklov Mathematical Institute of Russian Academy of Sciences, Fontanka 27, 191023, St. Petersburg, Russia
E-mail: yumat@pdmi.ras.ru

Zhi-Wei Sun*

School of Mathematics, Nanjing University, Nanjing 210093, People's Republic of China
E-mail: zwsun@nju.edu.cn

In this paper we show that there is no algorithm to decide whether an arbitrarily given polynomial equation $P(z_1, \ldots, z_{52}) = 0$ (with integer coefficients) over the Gaussian ring $\mathbb{Z}[i]$ is solvable.

Keywords: Hilber's Tenth Problem; Diophantine equation; Gaussian ring; undecidability.

1. Introduction

The original HTP (Hilbert's Tenth Problem) asks for an (effective) algorithm to test whether an arbitrary polynomial Diophantine equation with integer coefficients has solutions over the ring $\mathbb{Z}$ of the integers. This was finally solved by Yu. Matiyasevich[5] negatively in 1970 based on the work of M. Davis, H. Putnam and J. Robinson[2]. Z.-W. Sun[10] showed further that there is no algorithm that can decide for any given $P(x_1, \ldots, x_{11}) \in \mathbb{Z}[x_1, \ldots, x_{11}]$ whether the equation $P(x_1, \ldots, x_{11}) = 0$ has integer solutions.

Let K be a number field which is a finite extension of the field $\mathbb{Q}$ of rational numbers. It is natural to ask whether HTP over the ring O_K of algebraic integers in K is unsolvable. Clearly, if $\mathbb{Z}$ is Diophantine over O_K then HTP over O_K is undecidable with the aid of Matiyasevich's theorem. It is known that $\mathbb{Z}$ is Diophantine over O_K if $[K : \mathbb{Q}] = 2$ or K is totally

*Corresponding author

real (cf. J. Denef[3,4]), or $[K : \mathbb{Q}] \geqslant 3$ and K has exactly two nonreal embeddings into the field of complex numbers (cf. T. Pheidas[7]), or K is an abelian number field (cf. H. N. Shapiro and A. Shlapentokh[8]).

In this paper we study Diophantine equations with few unknowns over the Gaussian ring

$$\mathbb{Z}[i] = O_{\mathbb{Q}(i)} = \{a + bi : \ a, b \in \mathbb{Z}\}.$$

Our main results are as follows.

Theorem 1.1. *A number $z \in \mathbb{Z}[i]$ is a rational integer if and only if there are $v, w, x, y \in \mathbb{Z}[i]$ with $v \neq 0$ such that*

$$\begin{aligned}&(4(2v(2(2z+1)^2+1)-y)^2-3y^2-1)^2\\&+2(w^2-1-3y^2(2z+1-xy)^2)^2=0.\end{aligned} \tag{1}$$

Theorem 1.2. *For any r.e. (recursively enumerable) set $\mathcal{A} \subseteq \mathbb{N} = \{0, 1, 2, \ldots\}$, there is a polynomial $P(z_0, z_1, \ldots, z_{52})$ with integer coefficients such that for any $a \in \mathbb{N}$ we have*

$$a \in \mathcal{A} \iff P(a, z_1, \ldots, z_{52}) = 0 \textit{ for some } z_1, \ldots, z_{52} \in \mathbb{Z}[i]. \tag{2}$$

It is well known (cf. N. Cutland[1]) that there are nonrecursive r.e. subsets of $\mathbb{N}$. Thus Theorem 1.2 has the following consequence.

Corollary 1.1. *There is no algorithm to decide for any polynomial $P(z_1, \ldots, z_{52})$ with integer coefficients whether the equation*

$$P(z_1, \ldots, z_{52}) = 0$$

has solutions in $\mathbb{Z}[i]$.

We will provide some lemmas in the next section and then prove Theorems 1.1-1.2 in Section 3.

2. Some Lemmas

For $A, B \in \mathbb{Z}$, the Lucas sequence $(u_n(A, B))_{n\geqslant 0}$ is given by $u_0(A, B) = 0$, $u_1(A, B) = 1$, and

$$u_{n+1}(A, B) = Au_n(A, B) - Bu_{n-1}(A, B) \ \ (n = 1, 2, 3, \ldots).$$

Sun[9] studied arithmetic properties of such sequences as well as related Diophantine representations over $\mathbb{Z}$.

Lemma 2.1. *Let $A, B \in \mathbb{Z}$.*

(i) For any $k, n, r \in \mathbb{N}$, *we have the identity*

$$u_{kn+r}(A, B) = \sum_{j=0}^{n} \binom{n}{j} (u_{k+1}(A, B) - Au_k(A, B))^{n-j} u_k^j u_{j+r}.$$

(ii) Let $A, B, M \in \mathbb{Z}$ *with* $M \neq 0$. *Then* B *is relatively prime to* M *if and only if* $u_n(A, B) \equiv 0 \pmod{M}$ *and* $u_{n+1}(A, B) \equiv 1 \pmod{M}$ *for some* $n \in \mathbb{Z}^+ = \{1, 2, 3, \ldots\}$.

(iii) If $A > B \geqslant 0$, *then* $(A - B)^n \leqslant u_{n+1}(A, B) \leqslant A^n$ *for all* $n \in \mathbb{N}$.

Remark 2.1. Parts (i)-(iii) are Lemmas 2, 6, 8 of Sun[9].

Lemma 2.2. *Let* $A \in \{2, 3, \ldots\}$. *Then*

$$x^2 - Axy + y^2 = 1 \text{ with } x, y \in \mathbb{N} \text{ and } x \geqslant y$$

if and only if

$$x = u_{n+1}(A, 1) \text{ and } y = u_n(A, 1) \text{ for some } n \in \mathbb{N}.$$

Remark 2.2. This is a known result, see, e.g., Lemma 9 of Sun[9].

Lemma 2.3. *If* $x, y \in \mathbb{Z}[i]$ *and* $x^2 - 4xy + y^2 = 1$, *then* $x, y \in \mathbb{Z}$.

Remark 2.3. This follows from a more general result of Denef[3]; a proof for this particular case was also presented in Section 7.3 of Matiyasevich[6].

Lemma 2.4. *For* $x, y \in \mathbb{Z}[i]$, *we have*

$$x = 0 \wedge y = 0 \iff x^2 + 2y^2 = 0.$$

Proof. Though the result is known, here we provide a simple proof.

Suppose that $x^2 + 2y^2 = 0$ but $x \neq 0$ or $y \neq 0$. Then $xy \neq 0$ and $x/y \in \{\sqrt{2}\,i, -\sqrt{2}\,i\}$. As $x/y \in \mathbb{Q}(i) = \{r + si : r, s \in \mathbb{Q}\}$, and $\sqrt{2}$ is irrational, we obtain a contradiction. This ends the proof. □

Lemma 2.5. *An integer* m *is nonzero if and only if* $m = (2x + 1)(3y + 1)$ *for some* $x, y \in \mathbb{Z}$.

Remark 2.4. This is a useful observation of S.-P. Tung[11].

3. Proofs of Theorems 1.1 and 1.2

Proof of Theorem 1.1. (i) We first show the "if" direction.

Suppose that there are $v, w, x, y \in \mathbb{Z}[i]$ with $v \neq 0$ satisfying (1). In view of Lemma 2.4, we have

$$4(2v(2(2z+1)^2+1)-y)^2 - 3y^2 - 1 = 0 \tag{3}$$

and

$$w^2 - 1 - 3y^2(2z+1-xy)^2 = 0. \tag{4}$$

Let $y_* = 4v(2(2z+1)^2+1)$ and $w_* = w + 2(2z+1-xy)y$. Then

$$y_*^2 - 4y_*y + y^2 = (y_* - 2y)^2 - 3y^2 = 1$$

and

$$\begin{aligned} & w_*^2 - 4w_*y(2z+1-xy) + y^2(2z+1-xy)^2 \\ =& (w_* - 2y(2z+1-xy))^2 - 3y^2(2z+1-xy)^2 \\ =& w^2 - 3y^2(2z+1-xy)^2 = 1. \end{aligned}$$

Applying Lemma 2.3, we see that $y, y_*, w_*, y(2z+1-xy) \in \mathbb{Z}$. Thus both $2z+1-xy$ and w are rational integers.

Note that

$$\frac{|y_*|}{4} \geqslant 2|2z+1|^2 - 1 = |2z+1|(2|2z+1|-1) + (|2z+1|-1) \geqslant |2z+1|$$

and

$$(y - 2y_*)^2 = 3y_*^2 + 1 \leqslant 3y_*^2 + \frac{y_*^2}{16} = \left(\frac{7}{4}y_*\right)^2.$$

If $(y-2y_*)^2 = (\frac{7}{4}y_*)^2$, then we must have $|y_*|/4 = 1 = |2z+1|$, hence $z \in \{0, -1\}$ and $|y_*| = |12v| > 4$. Therefore

$$|y| > 2|y_*| - \frac{7}{4}|y_*| = \frac{|y_*|}{4} \geqslant |2z+1|.$$

Recall that $2z+1-xy \in \mathbb{Z}$, and write $x = a+bi$ with $a, b \in \mathbb{Z}$. Then

$$|y|^2 > |2z+1|^2 = |(2z+1-xy) + (a+bi)y|^2 = (2z+1-xy+ay)^2 + b^2y^2,$$

hence $b = 0$ and $x \in \mathbb{Z}$. Thus $2z+1 \in \mathbb{Z}$ and hence $z \in \mathbb{Z}$.

(ii) Below we show the "only if" direction. For $n \in \mathbb{N}$ we simply write u_n to denote $u_n(4,1)$.

Let $z \in \mathbb{Z}$ and $k = |2z+1|$. By Lemma 2.1(ii), for some $n \in \mathbb{N}$ we have $u_{n+1} \equiv 0 \pmod{4(2k^2+1)}$. In view of Lemma 2.1(iii), $u_{kn} \geqslant 3^{kn-1}$ and $u_{n+1} \geqslant 3^n$. Write $u_{n+1} = 4(2k^2+1)v$ with $v \in \mathbb{Z}^+$ and set $y = u_n$. Then

$$4(2v(2k^2+1) - y)^2 = (u_{n+1} - 2u_n)^2 = 3u_n^2 + 1 = 3y^2 + 1$$

with the aid of Lemma 2.2. By Lemma 2.1(i),

$$u_{nk} \equiv k(u_{n+1} - 4u_n)^{k-1} u_n \pmod{u_n^2}.$$

Let $q = u_{kn}/u_n \in \mathbb{Z}^+$. Then

$$q \equiv k u_{n+1}^{k-1} \equiv k \pmod{u_n}$$

since $k \equiv 1 \pmod 2$ and $u_{n+1}^2 = 1 - u_n^2 + 4u_n u_{n+1} \equiv 1 \pmod{u_n}$. Define $\varepsilon = 1$ if $z \geqslant 0$, and $\varepsilon = -1$ if $z < 0$. Then $\varepsilon u_{kn} = u_n(\varepsilon k - x u_n) = y(2z+1-xy)$ for some $x \in \mathbb{Z}$. Let $w_* = \varepsilon u_{kn+1}$ and $w = w_* - 2\varepsilon u_{kn}$. Then

$$w^2 - 3y^2(2z+1-xy)^2 = (u_{kn+1} - 2u_{kn})^2 - 3u_{kn}^2 = 1$$

by Lemma 2.2. Now it is clear that (1) holds.

In view of the above, we have completed the proof of Theorem 1.1. □

Remark 3.1. In view of Lemma 2.5 and the proof of Theorem 1.1, a number $z \in \mathbb{Z}[i]$ is an rational integer if and only if there are $s, t, w, x, y \in \mathbb{Z}[i]$ such that (1) holds with $v = (2s+1)(3t+1)$.

Proof of Theorem 1.2. By Theorem 1.1(ii) of Sun [10], there is a polynomial $f(z_0, \ldots, z_{10}) \in \mathbb{Z}[z_0, \ldots, z_{10}]$ such that $a \in \mathbb{N}$ belongs to $\mathcal{A}$ if and only if $f(a, z_1, \ldots, z_{10}) = 0$ for some $z_1, \ldots, z_{10} \in \mathbb{Z}$ with $z_{10} \neq 0$.

Let $F(v, w, x, y, z)$ denote the left-hand side of (1). For $z_k \in \mathbb{Z}[i]$, by Theorem 1.1, $z_k \in \mathbb{Z}$ if and only if $F(v_k, w_k, x_k, y_k, z_k) = 0$ for some $v_k, w_k, x_k, y_k \in \mathbb{Z}[i]$ with $v_k \neq 0$. Thus, $a \in \mathcal{A}$ if and only if there are

$$v_k, w_k, x_k, y_k, z_k \in \mathbb{Z}[i] \quad (k = 1, \ldots, 10)$$

with $F(v_k, w_k, x_k, y_k, z_k) = 0$ for all $k = 1, \ldots, 10$ such that $z_{10} \prod_{k=1}^{10} v_k \neq 0$ and $f(a, z_1, \ldots, z_{10}) = 0$. By the proof of Theorem 1.1, when $a \in \mathcal{A}$ we can actually choose $z_{10}, v_1, \ldots, v_{10} \in \mathbb{Z} \setminus \{0\}$ to meet the requirements. Therefore, in view of Lemma 2.5, $a \in \mathcal{A}$ if and only if there are

$$v_k, w_k, x_k, y_k, z_k \in \mathbb{Z}[i] \quad (k = 1, \ldots, 10)$$

such that $f(a, z_1, \ldots, z_{10}) = 0$, $F(v_k, w_k, x_k, y_k, z_k) = 0$ for all $k = 1, \ldots, 10$, and $z_{10} \prod_{k=1}^{10} v_k = (2s+1)(3t+1)$ for some $s, t \in \mathbb{Z}[i]$. Thus, in light of Lemma 2.4, (2) holds for some polynomial $P(z_0, z_1, \ldots, z_{52}) \in \mathbb{Z}[z_0, z_1, \ldots, z_{52}]$. This concludes the proof. □

Acknowledgments

We thank the referee for the helpful comments. This work is supported by the NSFC-RFBR Cooperation and Exchange Program (grants NSFC 11811530072 and RFBR 18-51-53020-GFEN-a). The second author is also supported by the Natural Science Foundation of China (grant no. 12371004).

References

1. N. Cutland, Computability, Cambridge Univ. Press, Cambridge, 1980.
2. M. Davis, H. Putnam and J. Robinson, *The decision problem for exponential diophantine equations*, Ann. of Math. **74** (1961), 425–436.
3. J. Denef, *Hilbert's Tenth Problem for quadratic rings*, Proc. Amer. Math. Soc. **48** (1975), 214–220.
4. J. Denef, *Diophantine sets of algebraic integers, II*, Trans. Amer. Math. Soc. **257** (1980), 227–236, 1980.
5. Yu. Matiyasevich, *Enumerable sets are diophantine*, Dokl. Akad. Nauk SSSR **191** (1970), 279–282; English translation with addendum, Soviet Math. Doklady **11** (1970), 354–357.
6. Yu. Matiyasevich, Hilbert's Tenth Problem, MIT Press, Cambridge, Massachusetts, 1993.
7. T. Pheidas, *Hilbert's Tenth Problem for a class of rings of algebraic integers*, Proc. Amer. Math. Soc. **104** (1988), 611–620.
8. H. N. Shapiro and A. Shlapentokh, *Diophantine relationships between algebraic number fields*, Comm. Pure Appl. Math. **42** (1989), 1113–1122.
9. Z.-W. Sun, *Reduction of unknowns in Diophantine representations*, Sci. China Ser. A **35** (1992), no. 3, 257–269. Available from the website `http://maths.nju.edu.cn/~zwsun/12d.pdf`
10. Z.-W. Sun, *Further results on Hilbert's Tenth Problem*, Sci. China Math. **64** (2021), 281–306.
11. S. P. Tung, *On weak number theories*, Japan. J. Math. (N.S.) **11** (1985), 203–232.

Author index

www.ingramcontent.com/pod-product-compliance
Ingram Content Group UK Ltd.
Pitfield, Milton Keynes, MK11 3LW, UK
UKHW022107170926
8624IPUK00024B/192

9 789819 815937